NATIONAL
ACADEMIES *Sciences*
Engineering
ACADEMIES *Medicine*

NATIONAL
ACADEMIES
PRESS
Washington, DC

Climate Change and Human Migration

An Earth Systems Science Perspective

———

Anne Frances Johnson,
Nancy D. Lamontagne,
Deborah Glickson, and
Lyly Luhachack, *Rapporteurs*

Division on Earth and Life Studies

I0109498

Proceedings of a Workshop

NATIONAL ACADEMIES PRESS 500 Fifth Street, NW Washington, DC 20001

This activity was supported by Grant No. 2022844 between the National Academy of Sciences and the National Science Foundation. Any opinions, findings, conclusions, or recommendations expressed in this publication do not necessarily reflect the views of any organization or agency that provided support for the project.

International Standard Book Number-13: 978-0-309-72528-6
International Standard Book Number-10: 0-309-72528-3
Digital Object Identifier: https://doi.org/10.17226/27930

This publication is available from the National Academies Press, 500 Fifth Street, NW, Keck 360, Washington, DC 20001; (800) 624-6242 or (202) 334-3313; http://www.nap.edu.

Suggested citation: National Academies of Sciences, Engineering, and Medicine. 2024. *Climate Change and Human Migration: An Earth Systems Science Perspective: Proceedings of a Workshop*. Washington, DC: The National Academies Press. https://doi.org/10.17226/27930.

The **National Academy of Sciences** was established in 1863 by an Act of Congress, signed by President Lincoln, as a private, nongovernmental institution to advise the nation on issues related to science and technology. Members are elected by their peers for outstanding contributions to research. Dr. Marcia McNutt is president.

The **National Academy of Engineering** was established in 1964 under the charter of the National Academy of Sciences to bring the practices of engineering to advising the nation. Members are elected by their peers for extraordinary contributions to engineering. Dr. John L. Anderson is president.

The **National Academy of Medicine** (formerly the Institute of Medicine) was established in 1970 under the charter of the National Academy of Sciences to advise the nation on medical and health issues. Members are elected by their peers for distinguished contributions to medicine and health. Dr. Victor J. Dzau is president.

The three Academies work together as the **National Academies of Sciences, Engineering, and Medicine** to provide independent, objective analysis and advice to the nation and conduct other activities to solve complex problems and inform public policy decisions. The National Academies also encourage education and research, recognize outstanding contributions to knowledge, and increase public understanding in matters of science, engineering, and medicine.

Learn more about the National Academies of Sciences, Engineering, and Medicine at **www.nationalacademies.org**.

Consensus Study Reports published by the National Academies of Sciences, Engineering, and Medicine document the evidence-based consensus on the study's statement of task by an authoring committee of experts. Reports typically include findings, conclusions, and recommendations based on information gathered by the committee and the committee's deliberations. Each report has been subjected to a rigorous and independent peer-review process and it represents the position of the National Academies on the statement of task.

Proceedings published by the National Academies of Sciences, Engineering, and Medicine chronicle the presentations and discussions at a workshop, symposium, or other event convened by the National Academies. The statements and opinions contained in proceedings are those of the participants and are not endorsed by other participants, the planning committee, or the National Academies.

Rapid Expert Consultations published by the National Academies of Sciences, Engineering, and Medicine are authored by subject-matter experts on narrowly focused topics that can be supported by a body of evidence. The discussions contained in rapid expert consultations are considered those of the authors and do not contain policy recommendations. Rapid expert consultations are reviewed by the institution before release.

For information about other products and activities of the National Academies, please visit www.nationalacademies.org/about/whatwedo.

ORGANIZING COMMITTEE FOR WORKSHOP ON CLIMATE CHANGE AND HUMAN MIGRATION: AN EARTH SYSTEMS SCIENCE PERSPECTIVE

AMIR AGHAKOUCHAK (*Chair*), University of California, Irvine
SOMAYEH DODGE, University of California, Santa Barbara
ELIZABETH (BETH) FUSSELL, Brown University
LORI HUNTER, University of Colorado Boulder
SHANNA N. McCLAIN, National Aeronautics and Space Administration
DIEGO PONS, University of Denver
DANIELLE N. POOLE, Yale School of Public Health
JACKIE QATALIÑA SCHAEFFER, Alaska Native Tribal Health Consortium
KILAPARTI (RAMA) RAMAKRISHNA, Woods Hole Oceanographic
  Institution

Study Staff

LYLY LUHACHACK, Program Officer
MARGO CORUM, Visiting Scientist (*until December 2023*)
DANIELLE WOODRING, Associate Program Officer (*until January 2024*)
MAYA FREY, Senior Program Assistant
DEBORAH GLICKSON, Board Director

Reviewers

This Proceedings of a Workshop was reviewed in draft form by individuals chosen for their diverse perspectives and technical expertise. The purpose of this independent review is to provide candid and critical comments that will assist the National Academies of Sciences, Engineering, and Medicine in making each published proceedings as sound as possible and to ensure that it meets the institutional standards for quality, objectivity, evidence, and responsiveness to the charge. The review comments and draft manuscript remain confidential to protect the integrity of the process.

We thank the following individuals for their review of this proceedings:

KATHARINE DONATO, Georgetown University
DIEGO PONS, University of Denver, Colorado; Colorado State
    University; and Universidad del Valle de Guatemala
BENJAMIN ZAITCHIK, Johns Hopkins University

Although the reviewers listed above provided many constructive comments and suggestions, they were not asked to endorse the content of the proceedings, nor did they see the final draft before its release. The review of this proceedings was overseen by **BARBARA ENTWISLE,** University of North Carolina at Chapel Hill. She was responsible for making certain that an independent examination of this proceedings was carried out in accordance with the standards of the National Academies and that all review comments were carefully considered. Responsibility for the final content rests entirely with the rapporteurs and the National Academies.

Contents

Acronyms and Abbreviations

BEMS   Bangladesh Environment and Migration Survey

CHIRRP   Confronting Hazards, Impacts, and Risks for a Resilient
              Planet

FEWS NET   Famine Early Warning Systems Network

LEO   Local Environmental Observatory (network)

NASA   National Aeronautics and Space Administration
NSF   National Science Foundation

RIM   Resilience Inference Measurement (model)
RISE   NSF Division of Research Innovations, Synergies, and
              Education

1

Introduction

As defined in the report *Next Generation Earth Systems Science at the National Science Foundation*, Earth systems science "aims to discover and integrate knowledge on the structure, nature, and scales of interactions among natural (e.g., physical, chemical, and biological) and social (e.g., cultural, socioeconomic, and geopolitical) processes" (NASEM 2022). Following the release of the report, the National Academies of Sciences, Engineering, and Medicine (the National Academies) hosted a series of workshops to further explore how an Earth systems science approach may be used to address various aspects of climate change. On March 18–19, 2024, the National Academies hosted the Workshop on Climate Change and Human Migration: An Earth Systems Science Perspective as part of the series.

The workshop, sponsored by the National Science Foundation (NSF), sought to explore climate change impacts and the consequent influence on human migration. Climate-related migration can be temporary or permanent, can involve internal displacement within countries or crossing international borders, and can involve a broad array of other direct and indirect drivers. To consider these complex issues, the workshop speakers discussed the data, methods, and research strategies relevant to understanding climate-related migration.

Workshop Planning Committee Chair Amir AghaKouchak (University of California, Irvine) welcomed participants and provided context on the workshop. The workshop is the third workshop in a series of follow-on activities stemming from the *Next Generation Earth Systems Science at the National Science Foundation* report, with previous events covering tipping points, cascading impacts, and interacting risks in the Earth system (NASEM 2024b) and climate intervention in an Earth systems science framework (NASEM 2024a).

Climate change and migration are two interconnected phenomena that are reshaping societies and ecosystems around the globe. The impacts of climate change are forcing communities to confront the reality of displacement and relocation, AghaKouchak said, adding that this is not merely something to consider for the future but a reality that millions of people are facing today. The workshop featured five sessions and a series of keynote talks focused on climate change and migration across a broad range of temporal and spatial scales, including migration in response to individual events such as hurricanes and droughts, as well as in response to more gradual changes, such as rising sea levels. The workshop brought together people with a diverse set of perspectives and expertise to strengthen the understanding of the complex dynamics involved in climate-related migration at local, regional, and global scales. Participants exchanged knowledge, examples, and experiences; discussed key gaps in existing models, tools, and data used by physical and social scientists working on climate and migration; and highlighted opportunities for advancing the field. This Proceedings of a Workshop summarizes the issues discussed and highlights observations and suggestions made during the workshop. It is intended to provide a factual summary of the workshop discussions based on recordings and slides and does not reflect consensus views of the workshop participants or the National Academies.

NATIONAL SCIENCE FOUNDATION PROGRAMS AND PERSPECTIVES

Alexandra Isern (NSF) framed the workshop with remarks on how the NSF is approaching issues at the intersection of climate change and human migration, and Laura Lautz (NSF) highlighted additional NSF programs focused on integrating social and environmental systems.

Isern said that the workshop discussions can help NSF to map the future research investments needed to advance understanding and predictive capabilities to develop more effective policies related to human migration. She described how the *Next Generation Earth Systems Science at the National Science Foundation* report (NASEM 2022) has helped NSF frame a series of funding activities focused on research, translation, and innovation that are closely linked to NSF's crosscutting agency theme of building a resilient planet. NSF is making significant investments in climate change research and clean energy technologies while creating an integrated approach to engage scientists and engineers across disciplines with the goal of advancing knowledge, empowering communities, and generating innovative technological solutions that address societal needs and create resilient communities.

NSF's Geoscience Directorate has also released several Dear Colleague Letters[1] to focus community attention on priority areas that NSF seeks to advance. These topics include understanding and predicting the compounding effects of hazards and extreme weather and climate events; research at the intersection of climate change and human health; development of new technologies to advance Earth systems science research; and science governance and consequences of carbon dioxide removal and solar radiation modification strategies. Another area is the discovery, characterization, extraction, and separation of critical minerals that are needed in many clean energy technologies, which will be important in creating a more resilient society, Isern noted.

Additions to NSF's investments in climate resiliency were also announced through the Directorate for Technology Innovation and Partnerships Regional Innovation Engine Program, Isern said. Of the 10 new Engines awarded, four are closely related to changing environmental systems. These include the Colorado-Wyoming Climate Resilience Engine,[2] the Great Lakes Water Innovation Engine,[3] the Southwest Sustainability Innovation Engine,[4] and the Advanced Agriculture Technology Engine,[5] which is in North Dakota.

To recognize the importance of Earth systems science in facilitating a holistic approach to studying the planet, NSF also established a new division in the Geosciences Directorate called the Division of Research Innovations, Synergies, and Education (RISE).[6] This division's goal is to foster transdisciplinary collaborations that engage the broader geoscience community, with a focus on Earth systems–level research. It currently has five key incubators: geoscience education and diversity, cyber infrastructure, geoscience innovation, global climate challenges, and synergistic activities within and external to NSF. These incubators are helping the agency build partnerships with traditional partners as well as with philanthropies, industry, and others, Isern said, adding that RISE will help to foster connections within the geoscience community to drive transformative discoveries,

---

[1] See https://www.nsf.gov/pubs/2024/nsf24022/nsf24022.jsp (accessed June 14, 2024).

[2] See https://new.nsf.gov/funding/initiatives/regional-innovation-engines/portfolio/colorado-wyoming-climate-resilience-engine (accessed June 14, 2024).

[3] See https://new.nsf.gov/funding/initiatives/regional-innovation-engines/portfolio/great-lakes-water-innovation-engine (accessed June 14, 2024).

[4] See https://new.nsf.gov/funding/initiatives/regional-innovation-engines/portfolio/southwest-sustainability-innovation-engine (accessed June 14, 2024).

[5] See https://new.nsf.gov/funding/initiatives/regional-innovation-engines/portfolio/north-dakota-advanced-agriculture-technology (accessed June 14, 2024).

[6] See https://www.nsf.gov/div/index.jsp?div=RISE (accessed June 24, 2024).

innovations in workforce development, and end-use–inspired solutions to address urgent Earth systems challenges.

Isern commented that policies focusing on climate resilience, sustainable development, and social justice are essential for mitigating the impacts of migration and supporting affected communities, but developing sound policies involves identifying research gaps and advancing predictive capabilities to better understand the impacts of changing environments on human migration. "We are seeing clear evidence that climate change is exacerbating environmental stressors such as water scarcity, extreme weather events; compounding environmental hazards, sea-level rise; and it is also affecting disease vectors," she said. "These changes are particularly impacting those that are least able to adapt, creating significant disparities in climate equity with vulnerable populations such as low-income communities and Indigenous peoples being disproportionately affected by climate-related displacement." She expressed her hope that the workshop could help inform investments that benefit society with the basic research needed to advance goals such as creating early warning systems, improving disaster preparedness, and developing resilient infrastructure to enhance local adaptive capacity.

Lautz discussed Confronting Hazards, Impacts, and Risks for a Resilient Planet (CHIRRP), a program led by NSF's Geosciences Directorate that addresses the agency's priority of building a resilient planet and also reflects the increasing emphasis on use-inspired or transdisciplinary research focused on solutions and major societal grand challenges.[7] CHIRRP is informed, in part, by the *Next Generation Earth Systems Science at the National Science Foundation* report, Lautz said, with a goal of advancing Earth systems science and drawing connections between the geosciences and social, behavioral, economic, and biological sciences, as well as engineering. The program supports community-driven research partnerships focused on Earth system hazards with the ultimate goal of developing practical solutions for communities to help reduce risk and increase social and ecological resilience.

Lautz said that the program is looking for projects that integrate the physical and social dimensions of the Earth system, focusing on actionable solutions that help communities reduce risk and increase resilience and include equitable community partnerships. For the first year, CHIRRP is supporting planning grants, conference grants, research coordination networks, and mechanisms that can be used to help develop partnerships with communities where they do not already exist, provide training for more effective community engagement, or catalyze ideas.

---

[7] See https://new.nsf.gov/funding/opportunities/confronting-hazards-impacts-risks-resilient-planet (accessed June 24, 2024).

WORKSHOP THEMES

Several key themes emerged throughout the workshop presentations and discussions. Many speakers highlighted the complexity of human migration processes, encompassing various situations such as forced migration in response to a crisis that makes it necessary, migration in absence of crisis, and community versus individual relocations. Additionally, speakers stressed that migration is not always an individual choice but can also be a collective and context-dependent phenomenon, highlighting its multifaceted nature. Speakers also examined a variety of measurement and modeling complexities that lead to significant challenges in understanding climate-related migration, and pointed to critical data gaps and sources of uncertainty that can further complicate efforts to grasp how climate change and other complex factors, beyond climate or environmental, interact to drive migration dynamics.

Several workshop participants underscored the importance of holistic approaches that integrate diverse perspectives and address both data and knowledge gaps to enhance understanding of climate-induced migration and advance a research agenda in this area. Many participants also highlighted the importance of incorporating both global and local perspectives, along with an understanding of the impact of policies on migration. Speakers also explored opportunities to engage communities and incorporate local and Indigenous sources of knowledge in scientific models and approaches to understand historical and cultural drivers of or barriers to migration and consider issues on equity, ethics, and justice. Finally, several participants emphasized the importance of interdisciplinary collaboration between the social sciences and Earth science as well as more inclusive and diverse approaches to address this complex issue.

2

Overview on Climate Change and Human Migration: Current Status and Research Gaps

Three keynote speakers set the stage for the workshop for a multi-faceted exploration of climate change and human migration. Speakers provided an overview of different approaches to studying the topic, shared their perspectives on the current state of knowledge, and highlighted gaps that could be further explored.

INTEGRATING AN EARTH SYSTEMS PERSPECTIVE

Justin Mankin (Dartmouth College) began the session by discussing the merits and challenges of viewing climate migration using an Earth systems perspective. He posited that adopting an Earth systems perspective on climate migration is valuable, but he cautioned that it is essential to rectify data gaps in order to avoid reinforcing overly reductive and deterministic models of human tragedy.

He noted that people are paying more attention to climate change as temperatures continue to break records at an accelerating pace and as the impacts of climate change grow increasingly apparent in daily life for many. For example, a person born in 1950 has now experienced more than a year's worth of daily temperature records, or around 365 new temperature records, while a person born today can expect to reach that threshold of a year's worth of additional records in the next 25 years. These changes come with staggering costs, he said. Based on econometric modeling, Mankin's team estimated that about $30 trillion worth of global macroeconomic income losses were due to extreme heat from 1992 to 2013 (Callahan and Mankin 2022). They also estimated that an El Niño event in the tropical

Pacific that began in March 2023 will be associated with losses on the order of $4 million to $5 trillion globally, disproportionately impacting the most vulnerable (Callahan and Mankin 2023). All of these impacts are happening amid a wider geopolitical context and social upheavals, which together drive people to move in order to manage their exposure and stress, Mankin said, noting that 1 in 75 people faced social disruptions that forced them to move in 2022, with more than half of those moves being internal displacements.

The value of using an Earth systems approach is that it considers the coupled dynamics of a system, and Mankin suggested that meaningfully incorporating people into models of the climate system can shed light on both future climate changes and the human impacts of those changes. He presented a framework that depicts how climate forcing leads to climate change, which induces a set of impacts that alter human well-being, which in turn induce a human response that feeds back into the process (Figure 2-1). Human responses to climate impacts can come in myriad forms and are projected onto complex socioeconomic and cultural landscapes, Mankin said, and he explained how this framework can help to elucidate the ways in which some responses may diminish exposure risks while others could produce feedbacks that further exacerbate climate change. To inform interventions aimed at managing climate change risks and understand the potential impacts of increased mobility and migration, he noted that it is important to recognize the extent to which warming influences those processes and vice versa.

Using this framework, Mankin and colleagues have traced the relationships between emissions and impacts, and results indicate that the countries that are suffering the greatest losses due to climate impacts are generally those that are the least responsible for global warming (Callahan and Mankin 2022b; Mankin and Callahan 2022). Pointing to the many equity and justice issues raised by the costs of climate change and adaptation, these findings underscore the importance of integrating people into the models through which we understand the climate system. However, there are also significant challenges in meaningfully integrating an Earth systems approach to climate migration. Disentangling causality includes reconciling different spatiotemporal scales, various types of uncertainty, and the different forms of prediction used across disciplines, Mankin said. He added that overcoming these challenges will involve figuring out the right balance between specificity and generalizability and determining what level of reduction is acceptable.

Another key challenge is that understanding the impacts of climate change necessitates both climate and social science data that are often unavailable or that come with significant uncertainties. The people and places that are the most vulnerable and exposed to climate impacts are

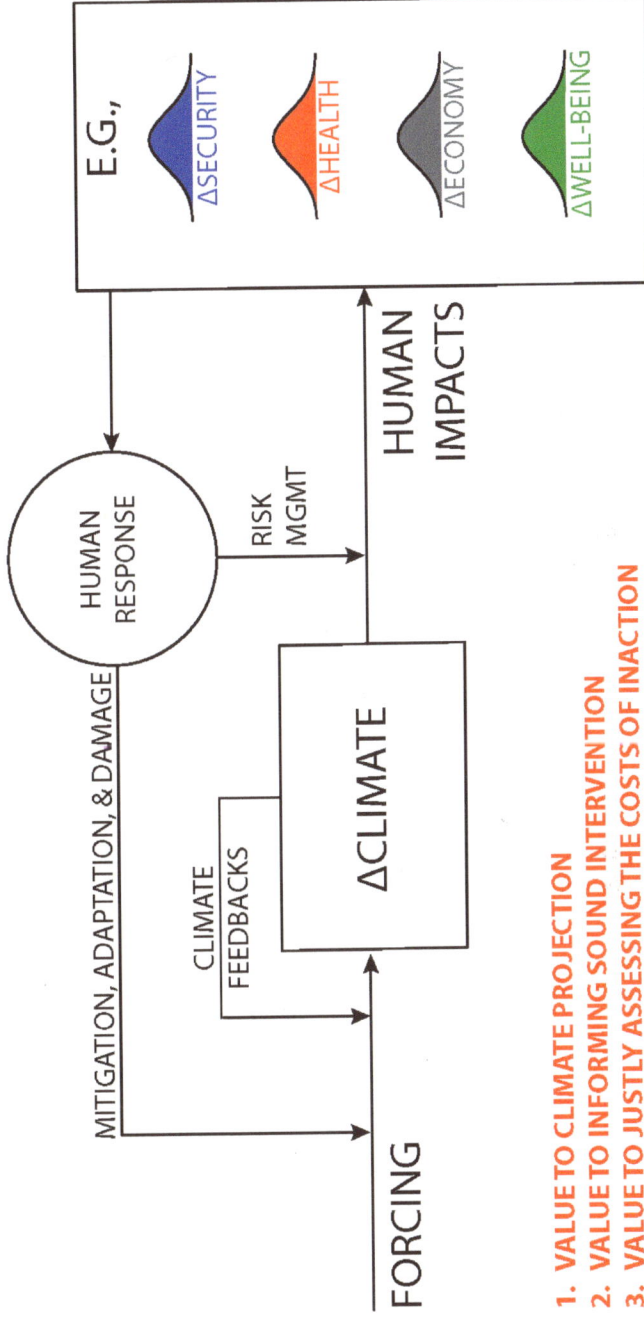

FIGURE 2-1 Endogenizing framework reflecting how human responses feed into the forcing mechanisms of climate change and the human impacts of that change.
SOURCE: Presented by Justin Mankin on March 18, 2024.

also the least likely to have reliable data on the relevant physical and social processes underway in their communities, Mankin said. This suggests that impacts and costs are likely being systematically underestimated, making it difficult to develop meaningful interventions to respond. Mankin stressed while this data poverty is widely recognized on the social side, it also creates important limitations on the physical side. For physical data, he noted that many highly populated but lower-resourced regions do not have any weather observations at all, making it difficult to track climate impacts in the places that are least culpable for them, experience them first, and have the fewest resources to manage them. For example, while North America and Europe are replete with sources of data on weather and climate, the same is not true for South America, Africa, parts of the Middle East, South Asia, and the interior of Asia. Mankin said that a lack of climate data in places experiencing climate change first is a critical need to address data inequities around the globe (Figure 2-2).

INCORPORATING SOCIAL SCIENCES

Lori Hunter (University of Colorado Boulder) discussed approaches to studying migration—a complex social process—from a social science perspective. She noted that a report published by the World Bank projected that 140 million people would move within their countries by 2050 due to climate change (Rigaud et al. 2018), underscoring the scale of anticipated climate-related migration. When thinking about the drivers of migration, Hunter said, it is important to recognize how micro-level drivers, such as age, education, wealth, and marital status, interact with meso-level factors, such as social networks and policies, to influence a person's decision to migrate or stay. These are also influenced by macro-level drivers, including political, demographic, and economic factors. Researchers captured some of these interactions in a framework that describes how environmental factors such as heat waves and drought interact with other processes to influence migration decision making (Black et al. 2011) (Figure 2-3). The complexity of the micro-, meso-, and macro-level factors at play in the context of climate-related migration makes it challenging to model migration decision making, Hunter said. She added that scale is also important to consider because the people making migration decisions are embedded in households, communities, and counties, and said that there is a critical need for accessible measures at different scales, pointing to a recent review that found many studies focused on precipitation or temperature more than other climate-related events (Hoffman et al. 2021).

Although modeling climate-related migration decision making presents a daunting challenge, various sources of data can offer useful insights. Hunter said that while there may generally be sufficient social data in terms of age, education, and gender of individuals as well as good environmental

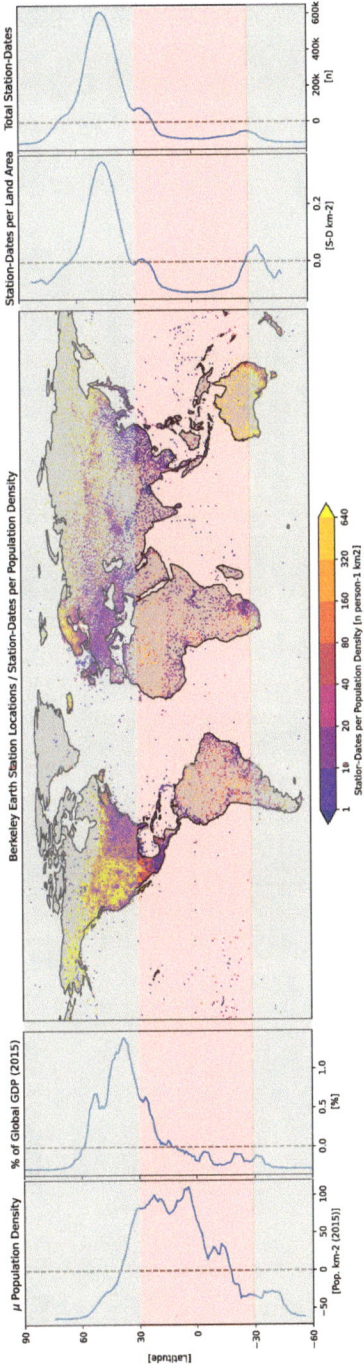

FIGURE 2-2 Distribution of climate data capabilities worldwide depicted as the number of weather observations given the underlying population density at a location.

NOTE: Global climate data collection is concentrated in North America, Europe, and Australia, while vast swaths of highly populous areas across Africa, Asia, the Middle East, and South America lack adequate climate data resources.

SOURCE: Presented by Justin Mankin on March 18, 2024.

12

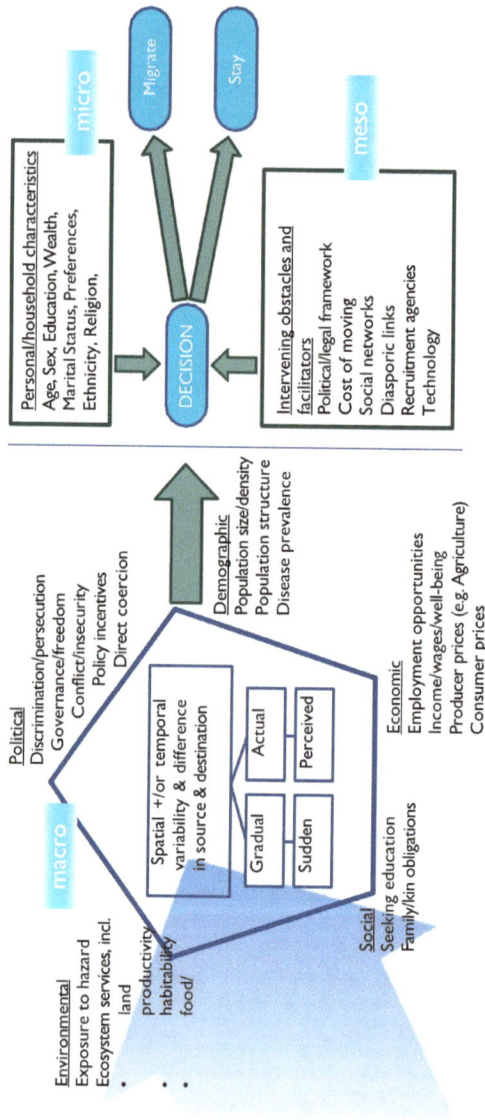

FIGURE 2-3 Conceptual framework describing micro-, meso-, and macro-level drivers of migration. SOURCES: Presented by Lori Hunter on March 18, 2024, from Black et al. (2011).

data on extreme events, temperature, and precipitation, the gap lies in bringing together those two kinds of data. For example, she noted that when people in the United States move across county boundaries, it is useful to know what extreme weather events have happened in that county, but this is only possible if data are available at the county level. The temporal scale is also important for understanding whether someone moves in response to an extreme event that happened last year or a series of events that have happened over the past 5 years.

Censuses can be a particularly useful data source for studying migration. For example, Hunter's team used data from the 2000 and 2010 Mexican censuses to study migration from Mexico to the United States. After linking these data with community-scale information, the researchers observed that migration out of drought-stressed regions of Mexico was prominent, but only for communities with strong social networks in the United States. However, Hunter said that censuses usually capture only whether someone has moved in the past 5 years, making it challenging to link with environmental data covering other time frames. Surveys offer another useful data source. For example, researchers used surveys to study seasonal and circular migration (capturing whether a household member migrates to earn income as well as whether those migrants come back) in a community in a small district in Thailand for more than 20 years (Entwisle et al. 2020). They found that climate-related stress in the communities did not affect out-migration but did make it less likely that individuals who had moved would come back. Ecological analysis, which involves characterizing different spatial units, can also be used to examine migration. For example, researchers predicted sea-level rise from climate models of counties across the United States and then examined what migration flows might be expected at the county scale (Hauer 2017).

Environmental pressures come in many forms that intertwine with economic, cultural, and other factors, and Hunter stressed that migration is just one way that households can respond to climate change. She suggested that it would be useful to have more information on other types of adaptation strategies. Additionally, while many studies examine what pushes people *out of* some areas, she said that more research is needed to understand what might attract them *into* other ones. She also noted that people who cannot move or choose not to move may be an understudied group and added that it is important to consider the impact of agency in determining a person's decision to move (or not), or whether they have no choice and thus no agency in the decision. She also suggested that it would be useful to improve methods for integrating qualitative insights, such as critical life and cross-generation experiences, with scientific knowledge. Finally, she suggested that fostering connections with policymakers could help ensure that the insights gained through studies of migration drivers can help to inform decision making.

CENTERING NATURE AND COMMUNITIES

Jacqualine Qataliña Schaeffer (Alaska Native Tribal Health Consortium) highlighted the value of centering nature and local communities and cultures in making decisions about climate migration. Introducing herself in the Iñupiaq language, she emphasized the importance of preserving Indigenous languages due to the unique concepts they convey, which is relevant not only in cultural and social contexts but in scientific research as well. She described how historical factors such as U.S. government actions have influenced climate vulnerability in Alaska, a vast and diverse area that is home to 229 federally recognized tribes. For example, she said that disruptions to traditional seasonal migrations have isolated some communities on fragile land. Highlighting the philosophical perspective offered by Indigenous elders, she said that maintaining balance entails understanding Earth's place in the universe, sharing the story of people's interconnections, and cultivating a harmonious relationship with nature.

"For 500 generations we have lived on the same geographic spot on the planet—we have not migrated," said Schaeffer. "We still eat our traditional seasonal foods that my ancestors ate. Why is this important? Because there is much to learn from people who have adapted over time. Even though our history is oral history, those stories are compelling when you cross them into science." She added that it is useful to consider the movement of people side by side with the movement of animals and other living organisms because they often presage change, underscoring the insights that can be gained from fostering connections with nature and the environment.

Of the 229 tribes in Alaska, Schaeffer said that 144 are environmentally threatened, and most likely every community in Alaska is being affected in some way by climate change. Alaska is warming three to four times faster than most other places on the planet, and the impacts are real, Schaeffer stressed. Yet, she lamented, many of the decisions affecting these communities are made elsewhere, by people who do not understand the environment, cultures, and interconnections between people and the environment that exist there. As an example, she pointed to the challenges faced in one village she visited that had no water and sanitation services and faced temperature extremes from –70°F (February) to 102°F (August). She noted that it is extremely difficult to engineer water and sanitation systems that work in these types of temperature extremes and that the inequities in Alaska are extreme and many communities are not even near the same baseline as many other Americans in terms of access to basic services. Additionally, she said that many people in Alaska do not want to move because they are on the lands of their ancestors, and their culture and value system emphasizes their role in stewarding that land and keeping those environments healthy.

The Alaska Native Tribal Health Consortium, a statewide tribal health network, examines environmental impacts through a tribal health lens. Schaeffer contrasted this with the Western approach to science and research, which often operates in silos without considering the psychological, mental, and behavioral health impacts of environmental pressures. She also noted that the recovery and response system in the United States is mostly reactionary when it comes to climate impacts—a mindset of waiting for disasters to happen rather than preparing for them. She suggested that for this perspective to change, there is a need for data from places such as the Arctic, where climate change is happening especially rapidly. The impacts in Alaska have been severe, including flooding, erosion, sea-level rise, and extreme weather events. As an example, Schaeffer pointed to a typhoon that washed away Indigenous fish camps that contained local communities' food resources for the year; she noted that policymakers who do not understand the way of life in these communities considered the fish camps to be recreational and thus were unprepared to adequately respond.

A report from the Alaska Native Tribal Health Consortium calls out challenges and inequities that are currently present across the system and provides recommendations on how to respond (ANTHC 2024). Schaeffer urged a focus on fostering curiosity to facilitate the connection and cross-pollination of different knowledge systems for greater understanding and impact moving forward. She said that having different worldviews does not preclude Indigenous communities and scientists from working together to support shared goals and the planet that we all treasure. Drawing from the wisdom of Indigenous elders—who are deeply knowledgeable about how ice systems affect broader ecological systems, including water levels, temperatures, and weather cycles—she emphasized the importance of understanding humanity's connections with our environment and closed with a message of hope that by embracing our shared history with nature, we can listen, learn, and adapt.

3

Mechanisms and Pathways for Modeling the Impacts of Catastrophic Events on Human Migration

In Session 1, speakers examined how rapid-onset catastrophic events such as severe storms may prompt human migration, including the interplay of natural and social processes under which stress-induced migration occurs as well as factors that can contribute to the resiliency of communities. Speakers from a range of disciplines discussed the mechanisms and pathways through which catastrophic events intersect with migration patterns and infrastructure.

FLOODS, DROUGHTS, AND HUMAN MOBILITY

Giuliano Di Baldassare (Uppsala University) spoke about the complex interactions and feedback mechanisms among hydrological, climatic, and social processes. He described research showing that climate-related migration mostly consists of temporary, local, and rural-to-urban mobility and discussed how water infrastructure can play a major role in influencing where people live and move.

Noting that climate-related migration rarely involves crossing national borders, Di Baldassare said that studying climate-related mobility is complex because it is often difficult to differentiate between drivers and to understand how agency factors into decision making (Schutte et al. 2021). The most vulnerable groups are the ones who have the fewest resources for moving and are therefore most likely to be trapped in place, he said. The empirical research, especially in the social sciences, shows that migration that takes place in response to extreme events such as intense flooding is often short-term local and/or rural-to-urban migration (IPCC 2014).

17

To study the relationship between climate-related events and changes in the spatial distribution of human settlements, Di Baldassare and colleagues use various types of survey analysis to examine local case studies and proxies to perform global studies of the distribution of human populations. For example, satellite images showing nighttime lights can be used to capture changes in human settlements over time, an approach the researchers used to study how the human settlement grew further from rivers over time (Mård et al. 2018). Countries experiencing the most flood fatalities, such as Mozambique, had the most increased distance of resettlement from the river over time as measured by the distance of the center-of-mass settlement from the river. This pattern of migration is less observed in countries with higher levels of structural flood protection (e.g., major levee systems), such as the Netherlands. In another study exploring the role of water infrastructure in mediating the link between floods, droughts, and human mobility in agro-pastoralist communities (Piemontese et al. 2024), the researchers found that when drought occurred, organizations try to cope by increasing the water supply. However, without effective governance, this tended to reduce mobility, an essential element of resiliency for pastoralist communities in response to climate change and drought.

When examining the link between climate change or climate-related extremes and human migration, Di Baldassare said that it is important to understand multiple perspectives because of the strong feedback mechanisms present. The complex interplay between dams, water supply, and population dynamics provides one example. Although water infrastructure is built to cope with growing population to meet increasing water demands, the presence of dams and increasing water supply itself is a driver of growing water demands. This increase in water supply and water demand has led to unsustainable water consumption, for instance, in the U.S. Southwest (Di Baldassarre et al. 2021). This feedback mechanism shows that the distribution of population and human mobility is not only a response to but also a driver of climate-related risk. In another study, researchers found that an increase in flood fatalities in Africa was primarily explained by population growth in flood-prone areas (Di Baldassarre et al. 2010), where in-migration increased people's risk from flooding but could also be a response to drought conditions, which has made settling close to the river more appealing.

During the discussion, Di Baldassare elaborated further on the relationship between policy, migration, and infrastructure. He emphasized the importance of research of studying these issues as coupled natural–human systems, which can expose the consequences of the interactions at play, such as increased vulnerability to drought and groundwater exploitation. This can help to reveal the risks and undesirable conditions arising from the legacy of water infrastructure and thus better inform decision making.

COMMUNITY RESILIENCE

Nina Lam (Louisiana State University) discussed climate change and human migration from the viewpoint of community resilience and mobility. "Human migration away from a place means that the place is not resilient," she said. "In the long term, resilience is sustainability." Describing her experiences studying community resilience and relocation in southeast Louisiana, an area that regularly experiences hurricanes and other severe events in addition to threats from rising sea levels, Lam emphasized that understanding disaster resilience begins with measuring it.

Lam and colleagues studied factors that influenced the return of small businesses after Hurricane Katrina, which, she noted, are important to local economies. They found that critical infrastructure protection—such as levee protections, utilities, and telecommunication—were the main factors involved in decisions to return. Lam noted that these same factors may impact human migration away from a place that has endured multiple climate-related events. In addition, the research suggested that emergency plans to help repair damaged property and rapidly restore infrastructure, resuming education and health services, and providing recovery aid to small businesses can help to improve social resilience and increase the likelihood that businesses will return after an event.

The Resilience Inference Measurement (RIM) model can be used to identify the social and environmental factors that make a community resilient. The model uses data on hazard threat, damage, and recovery to establish resilience scores and identify specific social and environmental capacities that contribute to community resilience. While these factors can be useful for examining the dynamics present in a community, Lam noted that getting the data on hazard threats is key to applying this method. The frequency or intensity of past natural hazards may be known, but predicting the location and frequency of future hazards is more challenging. To model indirect effects of environmental factors on population change with a system-dynamic model of migration in southeast Louisiana, Lam and colleagues extended from the RIM model to a dynamic resilience analysis using a Bayesian network (Lam et al. 2018). The results from a simulation using this model showed that the area closer to New Orleans would not be sustainable and would lose population in the next 50 years. The researchers also identified some hypothetical tipping points and possible mitigations that could affect this population loss; however, Lam noted that the accuracy of the models is constrained by limitations in data quality.

In another project, researchers surveyed more than 1,000 residents of southeast Louisiana for insights into how residents there think about moving. They found that 22 percent had considered moving, a relatively high proportion given that the average migration nationally is 11 percent. More than 25 percent of participants had experienced flooding from Hurricanes Katrina and Rita in 2005. Based on the survey, the researchers developed a decision model that showed how economics was actually more important than flooding in influencing relocation decisions in this vulnerable coastal area (Correll et al. 2021). In other words, Lam said, there must be economic opportunities available at the new location.

In some cases, an area may become uninhabitable, such as parts of coastal Louisiana due to land loss. For example, Lam pointed out that a Louisiana tribe became America's first climate refugees in 2016 when they had to relocate from Isle de Jean Charles due to the loss of 98 percent of their cultural land. In such extreme cases, as well as in situations when relocation is voluntary, Lam said that it is important to consider where people are moving to. Although migration to economically prosperous areas is desirable, people can also find themselves exposed to different risks, such as inland flooding, underscoring the complexity of balancing economic opportunities with environmental hazards. Lam added that the rising cost of flood insurance in Louisiana further exacerbates these issues, potentially leading to socioeconomic disparities in affected regions.

Lam ended by emphasizing that the broader context of resilience is crucial in addressing climate-induced migration. Questions about what constitutes community resilience and how to measure it remain pivotal. A comprehensive approach that acknowledges the interconnectedness of natural and human systems and captures the adaptive capacity of communities to bolster resilience is needed to address climate-induced migration, Lam said.

CHANGING HUMAN SETTLEMENT PATTERNS

Serena Ceola (University of Bologna) described how researchers can combine multiple types of data to shed light on climate extremes and changes in human settlement patterns. She highlighted examples of methods used to study the relationships between droughts and settlement patterns in Africa, along with insights gained through this research.

Droughts are among the most devastating hydrologic extremes in terms of both human lives and economic impacts, and climate change is increasing the likelihood and severity of droughts globally (AghaKouchak et al. 2014). An analysis of data from 1970 to 2019 showed that droughts represented only 7 percent of natural disasters during that period but were responsible for 34 percent of disaster-related fatalities (WMO 2021). Droughts in Africa have become particularly intense, frequent, and widespread over the past 50 years (IPCC 2023a; Masih et al. 2014).

As Mankin noted, short-distance mobility within national borders is the most prevalent form of migration (Caretta et al. 2022; Hoffmann et al.

2020; Xu and Famiglietti 2023). About half of the world's urban population is living in areas where migration has accelerated urban population growth (Niva et al. 2023), and disaster displacements are three times larger than displacements associated with conflicts and violence (Anzellini et al. 2020). Ceola noted that all climate change impacts, but especially hydrologic extremes such as floods and droughts, are affecting the risk that comes with migration, whether in the short term or permanently (Caretta et al. 2022).

A recent study examined associations between drought occurrences and human mobility toward urban areas in Africa between 1992 and 2013 (Ceola et al. 2023). For this work, the researchers used multiple data types, including data on annual drought occurrences from weather and emergency event databases, nighttime light analyses, geospatial data on the location of rivers, and urban population data from the World Bank. The results showed that droughts are influencing patterns of human settlements in Africa, with 74 percent of African countries seeing people move closer to rivers and 51 percent seeing people move closer to urban areas. Ceola said that one implication of this pattern of migration is that floods can cause more fatalities when they occur after a drought has driven people to move closer to the river. She also noted that some studies have predicted that the population living in Africa's drylands will double by 2050, which could set the stage for Africa's rapidly growing cities to become risk hotspots for both climate change and associated human displacements.

CREATING A MORE HOLISTIC VIEW OF RISK

Elizabeth Marino (Oregon State University Cascades) spoke about how meso-level factors such as property law and regulations can shape the way that risk is perceived and the way that planned and unplanned relocations happen in the wake of flooding disasters. She highlighted examples from a recent book featuring case studies from across the United States (Jerolleman et al. 2024) and a photo exhibit focused on places that are vulnerable to flooding and other disasters (Knowledge Is Power Exhibit n.d.).

She noted that there are complex social, political, and economic conditions that drive human migration and, as the impacts of climate change expand, resettlements linked to storms have become emblematic of climate change impacts. Marino posited that this new visibility around climate migration has created its own set of challenges, one of which is an oversimplified view of climate as a driver of migration. In Isle de Jean Charles, for example, she said, the history that has led to the current pattern of depopulation is more complex than sea-level rise. She noted that the Isle de Jean Charles is outside the levee system, leaving the island even more susceptible to storms. Disinvestment by the state and federal governments in protecting the road that provides access to the island made it difficult for people who needed reliable access to the mainland (e.g., for jobs or school). This led to a drop in population and meant that the island was even less likely to meet cost-benefit calculations for hazard mitigation measures, which may all be factors in migration.

It is critical to not only study whether people move or for how long and in what direction but also what their lives are like after the move occurs and why, Marino said. Relocation can result in an unbalanced distribution of risk and resilience, driving further inequity (Figure 3-1). She described a model that incorporates the policy landscape impact on people, communities, industrial activities, and development to understand risk exposures and opportunities after relocation. This model where climate impacts intersect with how laws and regulatory structures funnel risk into and away from certain areas may offer insight about historical and contemporary vulnerabilities, Marino said.

During the discussion, Marino highlighted additional complexities of relocation decisions, particularly in the context of environmental changes such as coastal shifts. She emphasized that while conventional analyses often treat relocation as a one-time decision aimed at reducing risk, people assess relocation differently, weighing the potential disruption to their livelihoods, family, and social ties against the promise of a better future. She suggested that there is a need to reframe the discussion to focus on maintaining cohesive livelihoods and sociocultural narratives during relocations, acknowledging the multifaceted nature of such decisions.

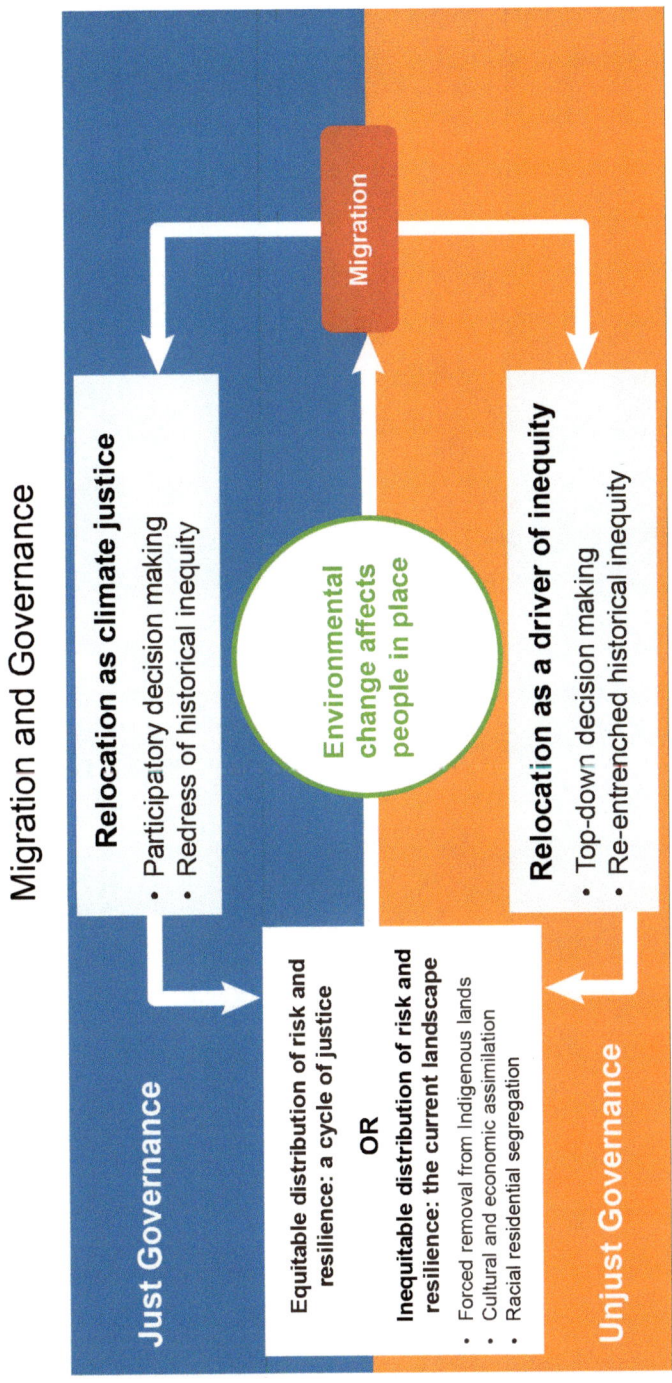

FIGURE 3-1 Schematic describing how just or unjust governance structures can redress or exacerbate inequities in the distribution of risk and resilience.
SOURCE: Presented by Elizabeth Marino on March 18, 2024.

4

Mechanisms and Pathways for Modeling the Impacts of Catastrophic Chronic and Slow-Onset Events on Human Migration

In addition to rapid-onset catastrophic events, Session 2 focused on understanding the drivers and impacts of chronic, gradual-onset, and sustained events such as droughts. Speakers examined the conditions under which these long-term shifts influence migration and examined some of the modeling and data challenges and the knowledge gaps in studying the complex interactions involved.

MODELING SLOW-ONSET CLIMATE EVENTS

Kevin Anchukaitis (University of Arizona) discussed the challenges of modeling a large system of interactions that link slow-onset climate events with food systems, socioeconomic factors, and the decision to migrate. His research is also examining how perceptions of these trends or changes may or may not lead to action.

Anchukaitis presented his work based in Central America in an area known as the dry corridor that is characterized by low rainfall and high levels of food insecurity and is one of the spots predicted for future drying. Various models and observational data suggest that in areas of expected climate signals, the change will emerge after four decades. Therefore, how chronic long-term changes and natural climate variability are managed, specifically average precipitation in this case and its subsequent effect on agriculture and food systems, will be important to consider, he said (Anderson et al. 2019).

One of the challenges in modeling is the insufficiency of data. Although there are weather stations in the region, precipitation datasets are not equally distributed or fully represented in time, Anchukaitis said. The data available for researchers may therefore come from only a few stations. Variability in data quality or what datasets are used in which models may result in different trends in precipitation. He noted that there is a lack of observations not just in Central America but also large parts of the tropics in the Global South and other regions where chronic climate conditions are expected to be observed. Climate models are useful tools, but it is also important to consider the potential limitations and uncertainties of the models as part of a larger system of downstream factors such as food security, said Anchukaitis.

Anchukaitis also discussed how people perceive changes in climate and how that perception can influence decisions about livelihood or migration beyond measurements and model simulations. For example, surveys of farmers in two Guatemalan communities on how they perceived change in rainfall over the last decade revealed that there was a wide range of responses, indicating that they perceived no change, less rain, or more rain. Perceptions also differed when comparing the two communities and led to different adaptive actions such as their agricultural practices. The clearest predictor of this difference was off-farm income; in a community heavily dependent on farming income, farmers were more apt to adjust their practices in response to perceived changes, whereas people in a community with more options for off-farm income perceived the same changes but were less likely to adjust their farming practices.[1]

In closing, Anchukaitis noted that climate models consistently suggest areas such as Central America and the Mediterranean will see significant drying in the coming century, and this is likely to impact vulnerable populations in these areas. Even as natural climate variability continues to be a dominant influence on agriculture, livelihoods, and food security, it is important to look ahead at these longer trends, how people in these regions perceive the climate changes they experience, and how they respond, he said.

CHALLENGES IN MODELING CLIMATE-INDUCED MIGRATION

Cristina Cattaneo (Foundation Euro-Mediterranean Center on Climate Change and RFF-CMCC European Institute on Economics and the Environment) examined some of the challenges in modeling relationships between climate change, weather shocks, and migration. She discussed how

---

[1] Presentation at the Workshop on Climate Change and Human Migration: An Earth Systems Science Perspective, March 18–19, 2024, see https://www.nationalacademies.org/event/41814_03-2024_workshop-on-climate-change-and-human-migration-an-earth-systems-science-perspective#sectionEventMaterials (accessed June 24, 2024).

key gaps in global migration data, limited knowledge of the interplay between alternative adaptation mechanisms and migration, and incomplete information about moderating factors or heterogeneous responses could hinder efforts to project future climate-induced migration flows.

One factor constraining the modeling of the relationships between climate change and international migration is the availability of high-quality migration data, she noted. Censuses and registries of destination countries are often used to track the nationality or place of birth of foreigners, and for many years these were the only data on international migration. Cattaneo said that while this approach allows large temporal coverage and provides some global perspective on origin and destination, it also has limitations. These data lack social demographic details, typically limited to only gender and education level and do not provide exact location within origin country. Detailed geographic and demographic data are important, as climatic conditions may widely vary within the same country.

Another source of data from origin countries is from surveys originally intended to assess conditions such as poverty and health. These surveys may provide geo-localized information on the place of origin and additional social demographic details. Cattaneo added that it can also be difficult to extract generalized messages and outcomes from data that come from a single country.

Newer sources of data, known as big data or digitally traced data, are available from social media with user-provided location data, Cattaneo noted. Information from mobile call records that capture cell tower location, or from remote sensing technology aboard satellites or drones can capture fine geographic detail, but Cattaneo noted that they tend to lack social demographic details and represent a limited time frame.

She emphasized that in addition to limitations in the data used to measure migration, there are also knowledge gaps that hinder efforts to model the links between climatic shocks and migration. Migration is one possible strategy for adapting to climate change, and it is often the last resort. However, it is not known what other options households and individuals consider in response to a climate shock and what makes migration more or less likely compared to other strategies. Cattaneo said that researchers lack a full understanding of the interplay between different adaptation strategies and climate shocks. Although there is a tendency to view climatic shocks as driving people to move, there are many contexts and circumstances in which people do not move, for example, because of lack of resources. For example, even if people have a high incentive to move, they may be less able to do so in situations where climate shocks reduce available resources. Consequently, it can be challenging to model migration responses to climate changes due to difficulty in determining whether immobility is due to lack of resources or to alternative strategies having been implemented.

Projections of future climate-induced flows are crucial modeling tasks, said Cattaneo, but there is a large knowledge gap in the available historical parameters that are needed for projections. Currently, most of the existing studies provide short-term dynamics, making it hard to predict migration patterns further in the future. Cattaneo noted two specific efforts to project future migration: the World Bank's Groundswell report (Rigaud et al. 2018), focused on internal migration, and research from Missirian and Schlenker (2017), exploring historical relationships and future flows of migration in specific destination countries.

DEFINING SLOW-ONSET CLIMATE IMPACT

Cascade Tuholske (Montana State University) discussed how nuances in defining slow-onset, or chronic, climate impacts can elucidate insights about climate migration mobility by connecting these impacts to socioecological systems. He offered that not having preconceived notions as to what might drive migration, a human action, in a complex socioecological system is important, pointing to a recent paper highlighting the inherent risk in climate or environmental determinism (Horton et al. 2021). The authors of the study suggested that ignoring human agency may oversimplify what is occurring on the ground in local communities. To understand the various dynamics at play for informed policy decision making, top-down modeling that can identify hotspots for decreased habitability and slow-onset events could be combined with local and social contextual information (Horton et al. 2021). Tuholske said that with the increasing availability of data, it will be important to work with transdisciplinary and cross-cultural teams to consider the context in any given location as to what may or may not be driving people to move.

"There is a serious risk of ecological fallacy whereby we observe movement of people often through secondary data, and then we associate it spatially with some sort of climate hazard or climate change event," Tuholske said. "Those two things may be correlated in space and time, but climate change is not, per se, the driver of movement of people. I really fear that we will over-attribute movement of people to climate change rather than to the social, political, or cultural contexts that might actually be creating underlying conditions where people don't feel safe or comfortable or are seeking new opportunity elsewhere."

Data limitations are another key issue, Tuholske said, pointing to a paper that examined the use of Earth observation data in areas that lack weather stations (Zaitchik and Tuholske 2021). Researchers estimated that 4 billion people live more than 25 kilometers away from a weather station with a reliable reporting record. In India, where there are approximately 3,000 urban settlements, there are only 111 weather stations. This makes it challenging to measure slow-onset climate changes in a given location and also influences modeling precision and accuracy. For example, data from the Coupled Model Intercomparison Project outputs offer resolutions only up to 100 by 100 kilometers, which does not capture the spatial variability in local climate, and most likely does not match up with the scale of location data for migration, Tuholske noted. However, researchers recently were able to achieve 5-by-5-kilometer outputs that can be applied to high-resolution climatology models (Williams et al. 2024). This improved resolution may then be used in a range of climate-related projections, such as for hot and dry extremes relevant to agricultural productivity and projecting the impacts of extreme heat on human health and labor productivity by approximated shaded wet globe temperature.

Moving forward, Tuholske emphasized the importance of trend-preserving higher-resolution climate projections and other analytical products that utilize multiple datasets. He added that it is important to elucidate the actual linkages among slow-onset events, human impacts, and decisions to migrate that are substantiated by actual experiences of people. "Conveying the complexity of the human decision to migrate is really paramount so we can really understand how climate change may be creating slow-onset events that drive people to leave their homes," he said.

DISCUSSION ON MODELING SLOW-ONSET EVENTS

Shanna McClain (National Aeronautics and Space Administration) moderated a panel discussion with the speakers that examined additional facets and limitations of data in modeling and understanding the impact of slow-onset climate events; perspectives on adaptation strategies and other factors that may influence the decision to migrate or stay; and the challenges of separating the influence of climate versus other factors in understanding why people move.

On resiliency, Tuholske noted that discussions on sustaining rural livelihoods and climate adaptation often consider agricultural technologies, particularly large-scale mechanized farming, as a solution to building resilience in farming communities and discourage migration. However, experiences from Europe and the United States suggest that increased reliance on machines can lead to decreased demand for agricultural labor over time and raise questions about whether technology truly enhances resiliency in rural communities facing climate change, if the aim is to provide alternative strategies to migration. While there is no definitive answer, he said that it is important to examine how new labor economic theories and historical development perspectives inform decisions regarding capital allocation to these communities, considering both their desires and the potential outcomes of such initiatives. Cattaneo added that migration is a last-resort choice; if climatic shocks can be addressed through adaptation, then people can choose to stay.

Alexander de Sherbinin (Columbia University) highlighted additional sources of data, including sentinel sites in Bangladesh, Mexico, and elsewhere, where researchers collect data across multiple years to try to understand migration and other demographic patterns. In addition, he pointed to long-term demographic surveys that collect information on household composition changes over time; have been conducted in some countries in Africa and other regions and can provide some information on youth mobility. Cattaneo added that relying only on country-level data requires estimating the specific challenges on an individual level. Ideally, she said, more data on both climate and information about where people are located would provide important information.

In response to a question about the impact of the general trend toward urbanization, Anchukaitis commented that urbanization trends also add to the complexity of modeling how climate changes might ripple through various communities. What are the consequences of increased urbanization and the climate change impacts in urban environments?

Cattaneo pointed out that people may sometimes be labeled as climate migrants even though other social and economic factors may be present. Climate has a direct role when, for example, a flood or sea-level rise occurs, and people are forced to move because their home is destroyed. However, it is more complicated for indirect relationships, which are more likely during slow-onset events. For example, climate impacts may affect economic factors, which in turn leads to relocation. Tuholske commented that the impact of an event within the current globalized economic system is location dependent. He emphasized that it is important to think about the interplay of various processes, what it means to be a climate migrant, and the consequences of formalizing that label.

5

Transdisciplinary Research and Collaborations

In this session, the speakers considered how transdisciplinary research is important to understand climate change impacts on migration from an Earth systems science approach and associated challenges in conducting said research.

CONSIDERATIONS FOR TRANSDISCIPLINARY RESEARCH

Diana Liverman (University of Arizona) offered a perspective on the nature, purpose, and practice of transdisciplinary research. Pointing to her own history of having migrated multiple times in response to job opportunities, family priorities, and the quest for adventure—but not for purposes of escaping climate change impacts—she cautioned against the temptation to attribute migration decisions to climate when many other factors are often at play, especially among people who have the resources to move out of choice rather than necessity.

Liverman sees transdisciplinary research as a collaboration that involves nonacademic partners who are co-producing knowledge with a commitment to action. She added that research on climate migration connects many disciplines and scientific methodologies, such as geographers, who are trained in both Earth science and social science, and utilizes various quantitative and qualitative methods and approaches ranging from modeling to ethnography. She noted that current work in climate and migration builds upon a long history of research in fields such as geography and anthropology. For example, she discussed how work in cultural ecology showed how people were able to creatively adapt to climate variability through social organization and technology. Geographers are using political ecology, which considers human agency, the physical environment, and political economy to understand climate–society relationships, and provide nuanced analysis of whether and why people move or stay. A long tradition of studying climatic and weather hazards can help to explore the linkages between climate and migration, particularly in understanding vulnerability to climate change, Liverman said.

"Sometimes it feels like this climate migration issue is very recent ... and it really is not," said Liverman. Research has shown many links between climate and mobility, including seasonal migrations, as a normal practice and not in response to crisis as well as adaptation to varying climates for decades. She suggested a need for on-the-ground investigations and rigorous experimental designs to better understand the complexity of climate and migration. Liverman stated that studies that rely only on statistical correlation do not provide a complete explanation because correlation is not causation and detailed research is needed to understand the explanation of statistical relationships, including the exploration of outliers and anomalies and confirmation through fieldwork.

From her work studying climate vulnerability and migration in Mexico, Liverman said she learned the importance of using multiple data sources, including local knowledge, to understand climate trends and impacts. She emphasized the value of considering both environmental and social history in community studies and the importance of engaging with communities before designing surveys on climate change by utilizing techniques such as the repertory grid—which asks respondents to compare elements of a landscape or community so that they construct what is important rather than use categories imposed by the interviewer. Another approach is to triangulate information by asking people about their neighbors' perceptions and decisions as well as their own.

Finally, Liverman added that truly transdisciplinary research requires careful thought about how researchers can give back to the community to avoid simply extracting information without providing any benefit to the people involved. For example, her colleagues have given materials to schools, paid locals to help with research tasks, and included local colleagues and nonacademic collaborators as authors on published articles.

CONNECTING THE PHYSICAL AND SOCIAL LANDSCAPES

Katharine Donato (Georgetown University) described a collaboration in Bangladesh as an example of transdisciplinary research that connects physical and social landscapes to examine how climate and environmental change affect patterns and processes of demographic outcomes, including migration. Prior to her involvement, she noted that a team of geologists, engineers, and Earth scientists had been looking at the physical landscape changes there for about 15 years, and she joined the team to provide the perspective of a social scientist. She also cited the National Science Foundation (NSF) Dynamics of Coupled Natural–Human Systems, now called the Dynamics of Integrated Socio-Environmental Systems,[1] program that supports interdisciplinary research to explore interactions of human and natural systems processes at diverse scales.

The study in Bangladesh focuses on water-related challenges and migration in the southwestern part of the country. River channel processes are undergoing significant shifts, and more change is projected in the coming decades, resulting in land loss and deposition on agricultural land, stress on flood embankments, more flooding, and increasing salinity, said Donato. She noted that migration has long been a common form of adaptation in this part of the world, as people are pushed away by various environmental stressors and pulled by various economic opportunities. Migration can be a sign of system failure, when people leave out of a sense of desperation to protect their families, but it can also result in remittances that ultimately strengthen families in communities. To study these dynamics, researchers collected migration histories that capture first and most recent migrant trips, including circular and return migration. Such data collected at detailed spatial and temporal scales permit integration with climate and physical science data and offer information about households that moved or stayed.

The first Bangladesh Environment and Migration Survey (BEMS or BEMS-1) was launched in 2013 and the follow-on BEMS-2 was done in 2019, yielding information on 4,000 households at 20 different sites reflecting a range of socioeconomic and environmental conditions (Carrico and Donato 2023). To track migration patterns, researchers used mixed methods by combining household surveys and remote sensing data to assess the relationship between land erosion and domestic or international migrant trips (Carrico and Donato 2019).

The researchers recently received additional NSF funding to conduct a longitudinal follow-up with about 1,500 households. For the next phase of the study, they plan to focus on the ways in which migration impacts the origin communities in terms of large-scale return migration and its impact on land use and well-being.

During the discussion, a workshop participant suggested some thought be given to the term "vulnerability," which is often taken to imply inherent characteristics of individuals. They proposed a shift toward viewing vulnerabilities as products of societal structures and inequalities. As an example,

---

[1] See https://new.nsf.gov/funding/opportunities/dynamics-integrated-socio-environmental-systems/nsf07-598/solicitation (accessed June 24, 2024).

the participant suggested that the link between heat events and early marriage that Donato noted could be attributed to the social system where girls are seen as economic burdens.

A PATH FORWARD FOR TRANSDISCIPLINARY RESEARCH

Pierre Gentine (Columbia University) spoke of the value of a transdisciplinary Earth systems approach to examine the interdependencies between migration and climate. He described how increasing pressures from extreme events, including flooding, heat stress, wildfires, droughts, and sea-level rise, are impacting livability, food productivity, and migration (Donat et al. 2016; IPCC 2023b; Vanos et al. 2023). He laid out a few factors for consideration in thinking about climate adaptation—accurate-prediction climate models, providing accessible data at scale, and capacity to adapt. Transdisciplinary collaboration is important to further understanding of these processes.

For example, Gentine noted that precipitation changes are not just occurring in the wettest regions of the world, but across the entire globe, adding that even if the annual mean precipitation changes only a little, the extremes of rainfall are also changing, leading to flooding events even in dry regions of the globe (Donat et al. 2016). Heat stress—which reflects a combination of temperature and humidity—is another factor that can contribute to large-scale migration and population movement. Models show that the number of days per year with dangerous heat is increasing, particularly in tropical regions with high population growth, potentially setting the stage for future large-scale across-country migrations (Sherwood and Huber 2010; Vanos et al. 2023; Vargas Zeppetello et al. 2022). Gentine emphasized that this is a complex issue and there are a lot of uncertainties, in terms of both climate modeling and accounting for other factors such as physiological acclimation to heat stress of certain populations. Transdisciplinary research is needed to truly understand the impacts of heat and heat stress, he said.

Droughts are also important because they impact food production, but Gentine said that in many regions, projections based on historical data provide a very uncertain view of the future of droughts. This can be a major issue because it affects planning for adaptations such as reservoirs or dams. Areas where there is more certainty as to the increasing intensity and frequency of droughts include Central America, western North America, and the Mediterranean, areas where Gentine noted that a lot of high-value crops are grown, which suggests that there could be substantial impacts on food productivity and food prices. Transdisciplinary research can provide a holistic view to understand the interplay of the impacts of droughts, food productivity, and migration.

Climate models have been extremely valuable tools, Gentine said, but still too uncertain to address many of the important questions around migration and climate change. Projections of even just global surface temperatures, which is one of the simplest climate metrics, show high levels of uncertainty because the physical system is so complex, and the uncertainties pose an even greater challenge when looking at finer scales. "When we think about adaptation [and] migration, it is not really a global metric that we want to consider," Gentine said. "The impact is at the very regional scale; we need to think of the regional level." At this level, climate models are even more uncertain, especially for extremes (Collins et al. 2011; Friedlingstein et al. 2014). Thus, he said, more work is needed to pinpoint climate events at a subcontinent or even more local level to understand the impacts that drive migration. In addition, Gentine underscored the need for more data from underserved communities, especially in regions such as the tropics, to understand and mitigate the impacts of climate change, and he emphasized that transdisciplinary research spanning disciplines from climate modeling to health, biology, sociology, and economics will be particularly valuable in this effort. Gentine noted that through funding by the NSF Science and Technology Center LEAP: Learning the Earth with Artificial intelligence and Physics,[2] he and colleagues are working to develop a platform that would provide a wide range of stakeholders with access to climate data in the cloud. He also noted that many factors affect the capacity to adapt.

---

[2] See https://new.nsf.gov/od/oia/ia/stc#active-centers-c98 (accessed June 24, 2024).

6

Regional Versus Global Perspectives for Modeling Climate Change–Related Migration Impacts

This session explored how evaluating climate change and migration at specific geographic locations affects the understanding and perception of migration flows and drivers. Speakers discussed how they incorporate geography at varying levels into their research and the value of incorporating various spatial analyses into study designs and research.

A GLOBAL LOOK AT MIGRATION

Paolo D'Odorico (University of California, Berkeley) highlighted insights from global studies on migration. Although migration is likely underreported in official records, he said, international migrants account for 2–3 percent of the global population according to census data. An analysis of the the global human migration network showed a pattern of increasing globalization in migration, reflected in a growing total number of migrants and an increasing level of interconnection among countries (Davis et al. 2013).

Overall, D'Odorico said that studies have revealed that the nexus between migration and environmental conditions can be extremely complex, with multiple causal factors and not driven solely by environmental conditions. There are also time lapses between environmental migration and its causes and drivers. He noted that where people go also depends partly on where they come from; the community structure of the global network of migrations shows modularity in that there is a community of countries within which people tend to move based on identities of language and religion.

A recent meta-analysis examined factors such as migration timing, volume, and geographic patterns in sub-Saharan Africa based on research in the environmental sciences and social sciences (Wolde et al. 2023). The study identified two major environmental drivers. One was excessive water events such as heavy rainfall, cyclones, and riverine flooding. At the other extreme was drought or water scarcity, causing poverty and famine. Each of these drivers was associated with the migration of about 6 million people.

Migration decisions are driven by complex arrays of pushes and pulls, direct and indirect factors, and environmental and non-environmental reasons. D'Odorico highlighted how radiation models can be used to examine the relationship between migration and environmental drivers by accounting for different pathways through which migrants are induced to move (Davis et al. 2018; Simini et al. 2012). This requires an analysis of the environmental drivers, how these drivers affect their livelihoods, and exposures to environmental risk and crop failure, he said.

HABITABILITY, SOCIAL DRIVERS, AND
CLIMATE MOBILITY MODELING

Alexander de Sherbinin (Columbia University) focused on the role of habitability in climate mobility modeling, the importance of social drivers, and methods for modeling migration. In 2021, de Sherbinin and colleagues defined habitability as the conditions in a particular setting that support healthy life, productive livelihoods, and sustainable, intergenerational development (Horton et al. 2021), and de Sherbinin noted that "at a reasonable cost" it could be added to this list. He said that climate change can undermine multiple dimensions of habitability, including basic survivability, livelihood, security, and society's capacity to manage environmental risk. "This has to do with essentially climate change undermining the very foundations of resilience," he said. "If your governance structures are being affected by climate impacts, if your GDP [gross domestic product] declines by 50 percent because of a storm, all of these things are going to undermine the capacity of society to respond."

Although top-down, process-based models driven by global climate models can help identify hotspots and warn of potential dire consequences, de Sherbinin cautioned that they are often overly deterministic. As other speakers noted, he said that distress migration will likely rise in response to climate stressors and some of this will be involuntary mobility. From a policy perspective, the Global Compact for Migration calls on governments to strengthen joint analysis and sharing of information to better map, understand, predict, and address migration (Global Compact for Migration n.d.). Toward this goal, de Sherbinin and colleagues proposed combining top-down assessments with bottom-up co-production of knowledge with local communities and stakeholders as well as bottom-up modeling approaches such as agent-based modeling, which are more suited to local circumstances (Horton et al. 2021).

There have been various hotspot mapping efforts based on climate factors such as sea-level rise, global temperature, heat stress, droughts, and floods (Giorgi 2006; Horton et al. 2018). It is also possible to produce maps using an ecological perspective by showing where humans tend to preferentially inhabit certain types of climatological niches and looking at climate change shifts that would change the optimal conditions for human habitability (Samson et al. 2011; Xu et al. 2020). Overlaying this with demographic pressure can reveal joint risks from future climate impacts and population issues.

Highlighting presentations from the March 2023 Population-Environment Research Network cyberseminar, "The Habitability Concept in the Field of Population-Environment Studies: Relevance and Research Implications,"[1] de Sherbinin underscored the multifaceted nature of habitability. Overall, researchers have described three dimensions of habitability: the collective ability to respond to risk (governance), livelihood resilience (e.g., food security), and physical and psychological safety. To understand habitability, he emphasized that it is important to consider both physical dimensions and socially constructed dimensions, which are often overlooked in research. For example, social dimensions can relate to capabilities, such as what people can do with the environmental resources they have or what power they have over their local environment. Emotive factors, such as a person's sense of attachment to the area they consider home is another social dimension. If populations are declining in an area, it may signal that habitability has declined, he added.

de Sherbinin stressed the importance of looking at the underlying social factors driving migration. As an example, he pointed to an analysis of migration in Senegal, which described how agricultural market structures have put smallholder farmers in a precarious financial position, leaving many indebted, with a limited economic future (Ribot et al. 2020). Although climate-related events may create the final push that drives some of these farmers to migrate, there are also other factors at play. To understand such issues, he added that it is key for academics in the West to consider the experiences of people categorized as migrants or displaced. He highlighted two studies examining different strategies for climate-migration modeling that does or does not incorporate migrants' experiences alongside environmental factors (Khosravi 2024; Tschakert and Neef 2022). For example, an examination of three surveys in West Africa found that the

---

[1] See https://populationenvironmentresearch.org/cyberseminars (accessed June 24, 2024); see also https://iussp.org/en/pern-cyberseminar-habitability-concept-field-population-environment-studies (accessed May 28, 2024).

top reason migrants cited for leaving their homes was financially related, followed by violence, family and personal reasons, and marriage and education. Environmental factors were less important. Even though environmental issues likely influenced some of the other factors cited, these findings demonstrate how the actual perceptions of migrants may differ from what scientists consider to be important drivers, de Sherbinin said.

Climate mobility models have varying capabilities, de Sherbinin said, noting that some are better at causal inference and other models are better in prediction. Modeling can be used at different temporal and spatial scales, ranging from local and very immediate scales of hours to days for evacuation modeling to longer-term migration trends. To use models effectively, de Sherbinin emphasized the need to ground modeling in migration theory. He noted that an analysis of about 75 climate migration papers found that the neoclassical theory (the basic argument that people move for rational reasons based on whether they can earn a higher wage somewhere else or due to wage differentials) dominated among the theories represented (de Sherbinin et al. 2022).

de Sherbinin summarized various modeling approaches that have been used to study climate migration and its impacts. For example, the exposure model focuses on the metrics or threshold of environmental factors such as sea-level rise or drought to determine when an area becomes uninhabitable. Agent-based models offer a bottom-up approach, exploring internal dynamics and policy implications, but require extensive data, de Sherbinin noted. Integrated assessment models, which are more complex and less common, analyze migration between countries. Finally, de Sherbinin noted that intercomparisons of model outputs to date show widely varying projections of migration patterns in response to climate change, emphasizing the nonlinear nature of these dynamics. He is working with colleagues at Columbia, Princeton, and Cornell Universities to develop a formal intercomparison process that would put climate mobility modeling on a more solid footing. Expressing agreement with other speakers at the workshop, he also drew attention to the need for more transdisciplinary collaborations. There are funding initiatives such as the NSF Global Convergence Research[2] program that support these collaborations. de Sherbinin also acknowledged that working across disciplines may be challenging and that there are barriers such as the lack of shared vocabulary and training and limited resources that need to be overcome.

---

[2] See https://new.nsf.gov/funding/opportunities/growing-convergence-research-gcr (accessed July 1, 2024).

DRAWING ON TRADITIONAL ECOLOGICAL KNOWLEDGE

Meda DeWitt (The Wilderness Society) discussed how sharing traditional ecological knowledge from Indigenous cultures can help guide communities in their response to climate change by sharing a story that has been recounted through thousands of years. During a time when the climate was changing rapidly, the Tlingit and Dene peoples who lived inland faced changes that disrupted traditional food sources and led to starvation, illness, and increased mortality. Recognizing the threat to their community's existence, they came together and decided to migrate in order to survive.

DeWitt relayed the story of their journey that included traveling through a river that went under a glacier. Two elder women decided to travel in a canoe under the glacier while two young men hiked over the glacier to map out the route and see if it was safe for their community. The community sang a grieving song to recognize the women who might perish and also the fact that their community was entirely dependent on their success. The women ultimately made it safely out to the coast and the rest of the community were able to follow the same route they had taken. They traveled to the coast, an area with an abundance of food, but grieved for their homeland. After a time, they adapted to living on the coast. DeWitt added that the community also has a song that helps repair grief and encourages people to enjoy living again.

DeWitt said that Indigenous people from all over Alaska have stories and traditional ecological knowledge relevant to climate change, often transmitted through stories like the one she shared. They know that Alaska was warm once and will be warm again. They have experienced the sea level rising and falling. While migration and climate change are not new experiences for them historically, she said, today's experiences are new due to geopolitical boundaries and policies that affect people's ability to migrate, and also because the human contribution has accelerated climate change. "Traditional ecological knowledge help[s] give us guidelines," she said. "We are challenged to grow, to evolve, to take responsibility for ourselves and our behavior, and to work together so that way we can thrive as we move through this."

During the discussion, it was noted that some communities are facing full relocation due to rising sea levels and erosion. DeWitt said that it is important to understand that the relocation process is not just about physical displacement but also about addressing longstanding inequities perpetuated by historical contracts and governance structures. These communities were historically forced to settle in vulnerable areas and have complex governance structures involving federal, tribal, and municipal entities. The fear of losing relationships, ties, and stories is a significant concern; DeWitt suggested that ideal solutions would involve moving entire communities together to preserve that sense of community.

Another problem is that funding for relocation comes from various sources, leading to fragmented efforts and long processes for relocation. DeWitt said that many Alaskan communities facing relocation due to climate change would welcome government assistance, but promises have fallen short. She added that communities feel both underresearched and overresearched, with little of the research benefiting them directly. It was also pointed out that various tribes have different cultures and different climates, so when developing solutions, it is important not to lump all tribal people together and to recognize that each tribe has its own history and knowledge. With increasing scientific recognition of climate change impacts and the value of traditional knowledge, DeWitt expressed hope for advancing collaborative solutions and a smoother transition. Schaeffer, the moderator for the session, similarly noted that it is challenging to make informed decisions when there is a lack of climate data and relationships with scientists and researchers.

7

Advancing an Earth Systems Science Approach to Modeling Climate Migration

The final workshop session built on the discussion of opportunities and challenges in advancing an Earth systems science approach to modeling climate migration. Speakers examined key gaps in data and modeling along with opportunities to overcome these gaps and advance the study of climate change and human migration.

HARMONIZING SPATIAL AND TEMPORAL DATA

Chris Funk (University of California, Santa Barbara) spoke about ways to harmonize spatial and temporal data and models to advance an Earth systems science approach to modeling climate migration. At the Climate Hazard Center, he has worked to co-develop datasets and strategies for monitoring extremes by collaborating with Earth scientists and data analysts from Africa, South Asia, and Central America to support the Famine Early Warning Systems Network (FEWS NET).[1] With funding from the U.S. Agency for International Development, FEWS NET monitors global food insecurity, which has increased from 35 million people who were extremely food insecure in 2015 to around 129 million in 2023.

Researchers at the Climate Hazard Center have developed strategies and datasets to support FEWS NET for about 25 years, and Funk said that much of this work is relevant to issues of migration. In particular, a staged early warning system framework for humanitarian crises related to drought illustrates how environmental factors can be approached at different levels

---

[1] See https://fews.net (accessed July 1, 2024).

of granularity and lead times to understand and anticipate pressures that people experience (Funk et al. 2023). Starting at a global scale and long lead times, climate forecasts are used to assess features such as El Niños and La Niñas. Then, at shorter lead times, interoperable weather forecasts are integrated to provide information at regional scales, high-resolution weather observations are used to track rainfall and temperature more locally, and finally, very high-resolution satellite imagery of vegetation can provide farm- and household-level information.

Funk pointed to several high-resolution datasets that are relevant for understanding climate impacts in the Global South. These include a dataset from thermal infrared geostationary satellite observations for 1983 to the present, which provides information about precipitation and temperature at local scales (Verdin et al. 2020). High-resolution heat projections are also available (Williams et al. 2024). In addition, the Climate Hazard Center's researchers perform high-resolution climatology studies, which are important for understanding where people are exposed to extreme precipitation or temperatures, as well as climate model reanalyses and projections. While each of these data sources has limitations, Funk said that they can be blended to create high-quality products that are widely used. For FEWS NET, researchers take information from global El Niños and La Niñas down to local weather, extreme precipitation, extreme temperatures, and precipitation deficits and translate this information into assumptions about whether people in particular locations are going to have enough food. This is an example of effective harmonization where Earth scientists from multiple agencies work closely with food security analysts and dynamic modelers to use Earth science information to anticipate future challenges, according to Funk.

Funk described some of the challenges researchers have encountered in complex modeling to be able to consider the dynamics of climate change impacts and migration from an Earth systems framework. For example, there are a number of variables to consider when looking at the impacts of temperature-related shocks in a warming world. Increasing climate volatility, warmer temperatures that can increase saturation vapor pressure (the amount of water air can hold at 100 percent humidity), and rainfall deficits that when combined with warmer air can draw more water from the stomates of leaves, can then cause plants to experience stress and irreversible damage. Impact on vegetation may then lead to downstream consequences related to food insecurity. Although this has received attention in data-rich areas such as the western United States, Funk said that relatively little focus has been given to it in the Global South. This makes it difficult to translate air temperature anomalies that occur, for example, over southern Africa, into quantitative impacts on food production and income. Overall, Funk said, the early warning community is much better at tracking the impacts of precipitation deficits than temperature extremes.

Another gap includes how to integrate evaluation of the impact of humid heat shocks on labor and health into warning systems. Funk said that there are research studies linking extreme humid heat to health impacts, reductions in labor productivity, and increased energy costs associated with air conditioners. However, there is little capacity to integrate this information into early warning systems. Furthermore, his new experimental data on wet-bulb globe temperature monitoring data show extreme wet-bulb globe temperatures over much of the planet, but there is not currently a way to rapidly incorporate the impacts of extremely humid and hot conditions on food security and livelihoods, Funk noted. On the opposite end of the spectrum, researchers are also working to understand risks associated with dry heat, which can be measured with vapor pressure deficits. Funk said that projections indicate a large increase in areas with either extreme humid heat risks or high vapor pressure deficits across the Sahel in Africa, parts of eastern Africa and, surprisingly, even in places in India by 2050 (Williams et al. 2024). This is predicted to lead to adverse health impacts, rapid plant senescence, longer dry seasons, and decreased crop production.

Overall, Funk said that carefully stacking multiple sources of information can help to overcome data scarcity, pointing to FEWS NET as a tangible illustration of the benefits of effective cross-disciplinary, cross-scale harmonization. However, more work may be needed in monitoring, anticipating, understanding, and mitigating temperature-related shocks. During the discussion, Funk suggested that the datasets compiled by FEWS NET could be valuable for informing policy making in the context of migration projections, noting that these datasets could be particularly useful in identifying emerging hazards, analyzing trends, and anticipating future risks. He also provided more details on stacking approaches to address the problem of poor high-resolution data or a lack of longitudinal data, describing as an example researchers' approach to assessing extreme wet-bulb globe temperatures in Guatemala, a country with significant variability at small scales. In this case, researchers combined historical station data with satellite observations and elevation data to create a high-quality climatology analysis and long-term average temperature for the region. Using satellite observations, climate model reanalyses, and temperature data, they then estimated anomalies around this average, providing a reasonable approximation of high-resolution variations despite not having as many observation stations as desired.

LEVERAGING LOCATION-BASED DATA

Ali Mostafavi (Texas A&M University) discussed how using location-based data can shed light on population movement and risk exposure (Hsu et al. 2024a). Location-based data are drawn from individual cell phones, which record the user's positions and movements as they move throughout the day (Fan et al. 2021). Over the past few years, this type of data has received a lot of attention from the scientific community because it can provide insights into behavior, especially during a crisis. During the COVID-19 pandemic, for example, this type of data was used to create empirical epidemiological models to evaluate viral spread and adherence to interventions such as shelter-at-home orders (Coleman et al. 2022; Fan et al. 2021, 2022). By revealing deviations and fluctuations in visitation and mobility patterns, location-based data can also be used to evaluate the impacts of climate-related hazards such as hurricanes and wildfires, in both the short term and the long term (Dargin et al. 2021).

For example, Mostafavi's group used location-based data to examine evacuation patterns in the area of Florida affected by Hurricane Ian in 2022. Using anonymized location-based data aggregated at the block group level to ensure privacy, the researchers were able to track where and when people evacuated ahead of the storm. The study showed how the near real-time availability of this type of data can produce timely insights during a crisis, as well as insights into longer-term changes in population as people relocate, either temporarily or permanently, in response to damage to their homes or neighborhoods. These data can also be associated with sociodemographic characteristics of communities such as income and racial characteristics to assess disparities in the ability to evacuate or relocate (Esmalian et al. 2022; Li and Mostafavi 2022). Mostafavi noted that empirical observational data on human activities do not always align with assimilation and model-based patterns. For example, research has shown that actual patterns of evacuation behavior are not consistent with any of the agent-based models reported in the literature, suggesting that incorporating location-based data might be useful to help validate or improve other types of models.

Mostafavi also described how location-based data can be used to understand populations' exposure to hazards between crises (Hsu et al. 2024b). For example, researchers used location-based data to evaluate where people go during their daily activities and how much of their time is spent in flood-prone areas (Lee et al. 2022). Defining a metric called life activity flood exposure, which is the proportion of minutes per week that a person spends in flood-prone areas, the researchers assessed flood exposure in different places across the United States (Rajput et al. 2024). They found that in San Francisco, for example, people spend only 6 percent of their time in flood-prone areas, while in Miami-Dade County, this proportion is 86 percent (Rajput et al. 2024). By accounting for where people spend their time and not just where they live, this approach provides more precise measurements of the degree to which populations are exposed to flood risks. The same approach can also be used to quantify exposure to other environmental hazards, such as air pollution, urban heat (Huang et al. 2023), and toxic sites (Liu and Mostafavi 2023), Mostafavi said, as well as capture changes over time as climate hazards and migration patterns evolve.

Mostafavi noted several challenges with using location-based data for longitudinal studies, including privacy protections, data limitations, policy variations, and bias and representativeness. Because of the anonymization of user data and changing user identities, Mostafavi said, it is difficult to track relocation patterns beyond about 6 months. He noted that over-coming this challenge would require agreements with data providers to ensure continued access to conduct longitudinal studies necessary to cap-ture population migration after climate-related events. Additionally, he suggested that establishing a global framework for collecting longitudinal data could enhance understanding of migration patterns and their inter-action with climate hazards on a global scale. Finally, to address issues of bias and representativeness (e.g., the fact that rural areas are likely to be underrepresented compared with urban areas), Mostafavi suggested that generative adversarial network models could be leveraged to create synthetic data and improve the utility of location-based data for analysis. "I think we have an opportunity for better, more near-time observational evaluation of human interplay with climate hazards and the impacts that climate change ha[s] on people," said Mostafavi. "But we need to have discussions, a framework, and collaborations between researchers, public organizations, and also private entities that collect and provide these data-sets so that we can have the opportunity to conduct longitudinal studies in a privacy-protected and, of course, ethical manner."

During the discussion, Mostafavi clarified that the researchers com-pared location-based evacuation data with agent-based models (ABMs) in the context of evacuation but not migration. The results showed that exist-ing behavior-based ABMs for evacuation did not accurately capture actual evacuation patterns; in light of this, Mostafavi suggested that combining location-based data with survey-based methods could improve ABMs, not-ing that each event is unique and requires precise measurements of behavior. He highlighted the importance of learning from errors in ABMs to better understand their limitations and improve future iterations.

DATA SOURCES FOR MIGRATION

Deborah Balk (City University of New York) discussed key data sources available for studying migration in relation to climate-related drivers, along with some considerations for using them effectively. She noted that al-though there are many uncertainties about the future, there are a few things that are known. One is that a greater share of the world's population will live in urban areas, and that this urbanization trend will be most acute in

Asia, Africa, and Latin America. In addition to increasing urbanization, she said that it is clear that in the future a greater percentage of the population will be older and that the world will be hotter, have more frequent and more intense storms and flooding, see more drought-prone areas, and experience higher sea levels.

The primary sources of demographic data available to understand current trends and project future ones include censuses, surveys, vital registration systems, tax records, basic demography (e.g., age, fertility, marriage, mortality) and health surveys. Though rich in many ways, there are limitations when used with climate and disaster data. Balk said that, in general, the spatial resolution of demographic data is somewhat coarse and often regional, which can pose problems for integrating demographic data with climate data. Differences in data formatting is another limitation in data integration. Finally, Balk stated that while datasets are often designed to be nationally or subnationally representative, they are usually not representative of environmentally delineated subgroupings that are most relevant to climate hazards, such as flood zones.

Climate data from satellites, Earth observations, and weather stations, also limit how they can be combined with other sources of information. She added that the format of satellite-derived information may be unfamiliar to social scientists working with demographic data. The spatial and temporal resolution of climate models varies from fine to coarse, resulting in a wide range of resolution.

As other speakers noted, Balk stated that internal migration within countries is generally more common than international migration and that more data are needed for internal measures. She suggested that improving assessment of internal migration may improve understanding of urbanization, and where people are moving to and from, particularly if this migration increases in response to climate-related stressors. Currently, there are relatively few studies comparing the prevalence of internal migration or urban migration across a large number of countries, making it difficult to know how countries compare in terms of internal migration. She noted that the IPUMS[2] data collection at the University of Minnesota allows for the harmonization of census and survey data but eliminates information that is not uniformly available, including origin destination information or reasons for migration. There are also varied sources of data for migration between and within countries, with different survey questions asked across data sources and over time.

Balk compared two approaches to obtaining migration data: censuses and demographic and health Surveys (Acosta et al. 2020). Censuses are fairly reliable for population projections because they fully enumerate the

_____
[2] See https://ipums.org (accessed July 1, 2024).

population, occur at fixed intervals, and capture longer periods of data. Demographic and Health Surveys capture population dynamics at a more granular level because they record local moves, can provide insights into flows between rural and urban areas, and are available for African countries, where historical censuses are largely absent.

Obtaining a global picture of subnational urban in-migration involves patching different types of data together. When doing so, Balk said it is important to be aware of differences between data sources in terms of what types of moves count as migration, potential sources of selection bias, the role of nonresponders, and differences in the bounding unit size. To improve migration statistics within national borders, Balk suggested focusing on migration between cities and towns, citing an example from Mexico showing that moves between major administrative areas miss a lot of moves within and to urban areas. This is important in the context of climate-related migration because evidence suggests that people tend to stay close, often within the same labor market, when relocating in response to disasters, Balk said, adding that such patterns can amplify inequities and have implications for who is in the path of the next possible storm.

Balk also highlighted emerging sources of data from cell phones, social media, and administrative data, such as Internal Revenue Service data relevant for moves within the United States, and vital registration data. For all of these sources, she said that it is important to think about who these data represent—only those with cell phones, only those who use Facebook, or only tax filers, for example. She suggested that combining such data sources in before/after studies or with other anchor data could enhance their utility. She added that partnerships are needed to aid in understanding these data and also stressed the importance of ensuring privacy protections.

Closing, Balk emphasized that measuring migration is critical but often more complex than measuring other demographic features. Even apart from climate change, future demographic shifts will be oriented around migration, so countries have an incentive to think hard about strategies to improve these measurements in their national statistical infrastructures.

DISCUSSION ON ADVANCING AN EARTH SYSTEMS APPROACH

During the panel discussion, speakers further examined challenges and opportunities related to data integration. They noted that data integration is important to advancing an Earth systems approach for understanding climate-related migration. Speaking about opportunities to bridge the gap between global, high-resolution climate data and location-based cell phone data, Funk noted that there are studies underway using cell phone data from Somalia and Zambia to examine where people are moving, which can be useful for food security monitoring. He added that Climate Hazards group Infrared Precipitation with Stations (CHIRPS) data have also been bundled with Demographic and Health Survey data to assess health impacts and suggested that there are many other opportunities for collaboration going forward.

Mostafavi noted that studies combining different datasets are bounded by the common denominator in terms of resolution. For example, if climate hazards or climate projections are at a certain scale, then movement data are typically aggregated at that same scale. This limits the ability to examine variations in climate hazards or compare different levels of exposure among populations. Another limitation is the lack of demographic information on individuals. Although this protection is necessary for individual privacy, it makes it difficult to directly analyze demographic characteristics, and researchers must rely on block group data as a proxy, which may not be as precise. He noted that having demographic information associated with the data would improve the resolution of findings.

Mostafavi also responded to a question about whether new sources of data could be used to penalize communities that fail to respond to government announcements regarding hazards, leading to increased insurance premiums or other penalties. He noted that this is a complex issue and there is a need to consider both risks and opportunities with the increasing availability and volume of different types of data. In addition to understanding potential risk, he said that such data could also be used to inform evacuation planning for future events.

Responding to a question about data sources relevant to understanding adaptation to specific hazards, Balk suggested that improving data collection methods to include both place of origin and destination could be useful. She noted that migration is just one possible response to hazards; other factors such as income or access to cooling resources may also play a role. She also emphasized the importance of studying individuals who do not migrate and suggested using longitudinal datasets to track their experiences over time, underscoring the need for systematic collection of data on adaptation policies and initiatives. Funk added that studying food insecurity can offer insights into understanding adaptation to hazards. Rather than directly modeling food insecurity, he said that researchers can focus on proximate drivers such as reductions in crop production and labor impacts. Similarly, when examining risks related to extreme heat exposure, it is important to investigate local labor opportunities, barriers, and income changes. To improve estimates of household adaptive capacity, he also noted that researchers are experimenting with using remote sensing to map poverty and adaptive capacity components at a high resolution. He suggested that advancements in all of these areas could inform future migration studies.

8

Closing Reflections

To close the workshop, members of the workshop organizing committee offered reflections on the common themes, challenges, and opportunities described during the presentations and discussions. Danielle Poole (Yale School of Public Health) noted that the workshop's intent was to bring together experts in a variety of disciplines and scholarship to explore how an Earth systems science perspective can advance the understanding of the impacts of climate change and human migration. A common theme that speakers pointed to is the importance of data and the data gaps, as well as approaches for data integration for a more holistic understanding of the different factors influencing migration, including but not limited to climate.

The workshop underscored how climate-related migration is an issue globally, Poole said. There are numerous stakeholders actively contributing to knowledge production related to climate migration models. A few presenters also noted the importance of engaging policymakers and model users to help researchers identify research priorities and questions. She also highlighted a few considerations in developing climate models and using the models to inform policies at local, national, and international levels including addressing model uncertainties, and risk communication. These discussions point to the importance of collaborative efforts in addressing the complex dynamics of climate migration.

Shanna McClain (National Aeronautics and Space Administration) highlighted some topics for further discussion. She noted that the workshop featured many conversations describing the complexity of modeling the different dimensions involved. More multidisciplinary research is important to better integrate the climate, environmental, and human dimensions.

McClain acknowledged that although there are challenges in overcoming research silos, there is increasing support for more collaborative approaches, including from federal research agencies. For example, she mentioned that the National Aeronautics and Space Administration (NASA) is working to roll out an Earth science-to-action strategy focused on societal benefit as part of advancing the scientific missions. McClain said that it is critical to sustain attention and financial support for transdisciplinary efforts that include a social science perspective, as well as the importance of incorporating Indigenous knowledge and human rights–based approaches that can provide key insights into what drives people to move, whether internally or internationally and whether temporarily or permanently.

She also suggested that it would be helpful to extend conversations about climate-related hazards and potential drivers of migration outside of academic spheres to include practitioners, lawmakers, and those involved with funding. As an example, she highlighted the recently launched NASA Lifelines program,[1] which is focused on bridging the gaps between Earth science and humanitarian action. "I'm hoping that the more we have efforts like this bringing science communities together with practitioner communities, the better changes we can make," she said.

Elizabeth Fussell (Brown University) reiterated the value of all the work in integrating climate exposure and human response data, while also recognizing existing challenges. She suggested that to inform future efforts, focusing on the key questions that need to be answered and the types of discoveries that would be most useful to address those questions would be useful. She added that social science data are essential to understanding opportunities to target resources toward those who are most at risk of being either displaced or locked in place.

To illustrate some of the complexities of data availability and use in understanding climate-related migration, Fussell highlighted research on the migration response after Hurricane Maria hit Puerto Rico in 2017. While many sources of data can shed light on this disaster and its aftermath—including data from social media, transportation statistics, consumer records, administrative data, surveys, and more—she said that it is important to consider not only what questions each type of data could be used to answer but also who could access the data and the time period covered. A study of population change due to outmigration after the hurricane showed the significance of various data types in understanding the magnitude and timing of the effect (Acosta et al. 2020). Fussell emphasized the importance of considering both short- and long-term perspectives in hazard analysis, and also highlighted how complex historical and socioeconomic factors may influence the impacts of climate crises and subsequent migration responses.

---

[1] See https://nasalifelines.org (accessed July 1, 2024).

To conclude, Fussell highlighted a resource that was shared with her during the workshop, called the Local Environmental Observatory (LEO) Network,[2] LEO Network is a platform to share environmental-related events or observations. In Kotzebue, Alaska, a resident reported on LEO that a lake near her house had suddenly drained, which led to exchange of knowledge between residents and scientists to figure out what had caused the lake to drain. It was ultimately determined to be the result of a complex chain of environmental causes, including the activity of beavers, among other factors. Fussell said that this illustrates a point made by several speakers on how partnerships between scientists and local community members can reveal the complex relationships between environmental changes and habitability of places.

---

[2] See https://www.leonetwork.org (accessed July 1, 2024).

References

Acosta, R. J., N. Kishore, R. A. Irizarry, and C. O. Buckee. 2020. Quantifying the dynamics of migration after Hurricane Maria in Puerto Rico. *Proceedings of the National Academy of Sciences* 117(51):32772–32778. https://doi.org/10.1073/pnas.2001671117.

AghaKouchak, A., D. Feldman, M. J. Stewardson, J.-D. Saphores, S. Grant, and B. Sanders. 2014. Australia's drought: Lessons for California. *Science* 343(6178):1430–1431. https://doi.org/10.1126/science.343.6178.1430.

Anderson, C. M., R. S. DeFries, R. Litterman, P. A. Matson, D. C. Nepstad, S. Pacala, W. H. Schlesinger, M. R. Shaw, P. Smith, C. Weber, and C. B. Field. 2019. Natural climate solutions are not enough. *Science* 363(6430):933–934. https://doi.org/10.1126/science.aaw2741.

ANTHC (Alaska Native Tribal Health Consortium). 2024. *Unmet Needs of Environmentally Threatened Alaska Native Villages: Assessment and Recommendations.* https://www.anthc.org/wp-content/uploads/2024/01/Unmet_Needs_Report_22JAN24.pdf (accessed June 24, 2024).

Anzellini, V., B. Desai, and C. Leduc. 2020. *2020 Global Report on Internal Displacement (GRID).* Internal Displacement Monitoring Centre, Norwegian Refugee Council. https://www.internal-displacement.org/publications/2020-global-report-on-internal-displacement-grid (accessed June 24, 2024).

Black, R., W. N. Adger, N. W. Arnell, S. Dercon, A. Geddes, and D. Thomas. 2011. The effect of environmental change on human migration. *Global Environmental Change* 21(December):S3–S11. https://doi.org/10.1016/j.gloenvcha.2011.10.001.

Callahan, C. W., and J. S. Mankin. 2022a. Globally unequal effect of extreme heat on economic growth. *Science Advances* 8(43):eadd3726. https://doi.org/10.1126/sciadv.add3726.

Callahan, C. W., and J. S. Mankin. 2022b. National attribution of historical climate damages. *Climatic Change* 172(3–4):40. https://doi.org/10.1007/s10584-022-03387-y.

Callahan, C. W., and J. S. Mankin. 2023. Persistent effect of El Niño on global economic growth. *Science* 380(6649):1064–1069. https://doi.org/10.1126/science.adf2983.

Caretta, M. A., A. Mukherji, M. Arfanuzzaman, R. A. Betts, A. Gelfan, Y. Hirabayashi, T. K. Lissner, J. Liu, E. Lopez Gunn, R. Morgan, S. Mwanga, and S. Supratid. 2022. Water. Pp. 551–712 in *Climate Change 2022: Impacts, Adaptation and Vulnerability.* Contribution of Working Group II to the Sixth Assessment Report of the Intergovernmental Panel on Climate Change, H.-O. Pörtner, D. C. Roberts, M. Tignor, E. S. Poloczanska, K. Mintenbeck, A. Alegría, M. Craig, S. Langsdorf, S. Löschke, V. Möller, A. Okem, and B. Rama (eds.). Cambridge, UK, and New York: Cambridge University Press.

Carrico, A., and K. Donato. 2019. Extreme weather and migration: evidence from Bangladesh. *Population and Environment* 41(September):1–31. https://doi.org/10.1007/s11111-019-00322-9.

Carrico, A., and K. M. Donato. 2023. Bangladesh Environment and Migration Survey (BEMS), 2019. Inter-university Consortium for Political and Social Research [distributor].

Ceola, S., J. Mård, and G. Di Baldassarre. 2023. Drought and human mobility in Africa. *Earth's Future* 11(12):e2023EF003510. https://doi.org/10.1029/2023EF003510.

Coleman, N., X. Gao, J. DeLeon, and A. Mostafavi. 2022. Human activity and mobility data reveal disparities in exposure risk reduction indicators among socially vulnerable populations during COVID-19 for five U.S. metropolitan cities. *Scientific Reports* 12(1):15814. https://doi.org/10.1038/s41598-022-18857-7.

Collins, J. A., E. Schefuß, D. Heslop, S. Mulitza, M. Prange, M. Zabel, R. Tjallingii, T. Dokken, E. Huang, A. Mackensen, M. Schulz, J. Tian, M. Zarriess, and G. Wefer. 2011. Major element data for tropical western Africa for samples from the Last Glacial Maximum, Heinrich Stadial 1, the Mid-Holocene and the Late Holocene. Supplement to Collins, J. A., et al. 2011. Interhemispheric symmetry of tropical African rainbelt over the past 23,000 years. *Nature Geoscience* 4(1):42–45. https://doi.org/10.1038/Ngeo1039. PANGAEA dataset publication series.

Correll, R. M., N. S. N. Lam, V. V. Mihunov, L. Zou, and H. Cai. 2021. Economics over risk: Flooding is not the only driving factor of migration considerations on a vulnerable coast. *Annals of the American Association of Geographers* 111(1):300–315. https://doi.org/10.1080/24694452.2020.1766409.

Dargin, J. S., Q. Li, G. Jawer, X. Xiao, and A. Mostafavi. 2021. Compound hazards: An examination of how hurricane protective actions could increase transmission risk of COVID-19. *International Journal of Disaster Risk Reduction* 65:102560. https://doi.org/https://doi.org/10.1016/j.ijdrr.2021.102560.

Davis, K. F., P. D'Odorico, F. Laio, and L. Ridolfi. 2013. Global spatio-temporal patterns in human migration: A complex network perspective. *PLoS ONE* 8(1):e53723. https://doi.org/10.1371/journal.pone.0053723.

Davis, K. F., A. Bhattachan, P. D'Odorico, and S. Suweis. 2018. A universal model for predicting human migration under climate change: Examining future sea level rise in Bangladesh. *Environmental Research Letters* 13(6):064030. https://doi.org/10.1088/1748-9326/aac4d4.

de Sherbinin, A., K. Grace, S. McDermid, K. van der Geest, M. J. Puma, and A. Bell. 2022. Migration theory in climate mobility research. *Frontiers in Climate* 4(May). https://doi.org/10.3389/fclim.2022.882343.

Di Baldassarre, G., A. Montanari, H. Lins, D. Koutsoyiannis, L. Brandimarte, and G. Blöschl. 2010. Flood fatalities in Africa: From diagnosis to mitigation. *Geophysical Research Letters* 37(22):2010GL045467. https://doi.org/10.1029/2010GL045467.

Di Baldassarre, G., M. Mazzoleni, and M. Rusca. 2021. The legacy of large dams in the United States. *Ambio* 50(10):1798–1808. https://doi.org/10.1007/s13280-021-01533-x.

Donat, M. G., A. L. Lowry, L. V. Alexander, P. A. O'Gorman, and N. Maher. 2016. More extreme precipitation in the world's dry and wet regions. *Nature Climate Change* 6(5):508–513. https://doi.org/10.1038/nclimate2941.

Entwisle, B., A. Verdery, and N. Williams. 2020. Climate change and migration: New insights from a dynamic model of out-migration and return migration. *American Journal of Sociology* 125(6):1469–1512. https://doi.org/10.1086/709463.

Esmalian, A., N. Coleman, F. Yuan, X. Xiao, and A. Mostafavi. 2022. Characterizing equitable access to grocery stores during disasters using location-based data. *Scientific Reports* 12(1):20203. https://doi.org/10.1038/s41598-022-23532-y.

Fan, C., S. Lee, Y. Yang, B. Oztekin, Q. Li, and A. Mostafavi. 2021. Effects of population co-location reduction on cross-county transmission risk of COVID-19 in the United States. *Applied Network Science* 6(1):14. https://doi.org/10.1007/s41109-021-00361-y.

Fan, C., X. Jiang, R. Lee, and A. Mostafavi. 2022. Data-driven contact network models of COVID-19 reveal trade-offs between costs and infections for optimal local containment policies. *Cities* 128:103805. https://doi.org/10.1016/j.cities.2022.103805.

Friedlingstein, P., M. Meinshausen, V. K. Arora, C. D. Jones, Al. Anav, S. K. Liddicoat, and R. Knutti. 2014. Uncertainties in CMIP5 climate projections due to carbon cycle feedbacks. *Journal of Climate* 27(2):511–526. https://doi.org/10.1175/JCLI-D-12-00579.1.

Funk, C., A. H. Fink, L. Harrison, Z. Segele, H. S. Endris, G. Galu, D. Korecha, and S. Nicholson. 2023. Frequent but predictable droughts in East Africa driven by a Walker circulation intensification. *Earth's Future* 11(11):e2022EF003454. https://doi.org/10.1029/2022EF003454.

Giorgi, F. 2006. Climate change hot-spots. *Geophysical Research Letters* 33(8). https://doi.org/10.1029/2006GL025734.

Global Compact for Migration. n.d. United Nations International Organization for Migration. https://www.iom.int/global-compact-migration (accessed May 28, 2024).

Hauer, M. E. 2017. Migration induced by sea-level rise could reshape the US population landscape. *Nature Climate Change* 7(5):321–325. https://doi.org/10.1038/nclimate3271.

Hoffmann, R., A. Dimitrova, R. Muttarak, J. C. Cuaresma, and J. Peisker. 2020. A meta-analysis of country-level studies on environmental change and migration. *Nature Climate Change* 10(10):904–912. https://doi.org/10.1038/s41558-020-0898-6.

Hoffmann, R., B. Šedová, and K. Vinke. 2021. Improving the evidence base: A methodological review of the quantitative climate migration literature. *Global Environmental Change* 71(November):102367. https://doi.org/10.1016/j.gloenvcha.2021.102367.

Horton, B. P., R. E. Kopp, A. J. Garner, C. C. Hay, N. S. Khan, K. Roy, and Ti. A. Shaw. 2018. Mapping sea-level change in time, space, and probability. *Annual Review of Environment and Resources* 43:481–521. https://doi.org/10.1146/annurev-environ-102017-025826.

Horton, R. M., A. de Sherbinin, D. Wrathall, and M. Oppenheimer. 2021. Assessing human habitability and migration. *Science* 372(6548):1279–1283. https://doi.org/10.1126/science.abi8603.

Hsu, C.-W., C. Liu, Z. Liu, and A. Mostafavi. 2024a. Unraveling extreme weather impacts on air transportation and passenger delays using location-based data. *Data Science for Transportation* 6(2):9. https://doi.org/10.1007/s42421-024-00094-1.

Hsu, C.-W., C.-F. Liu, L. Stearns, S. Brody, and A. Mostafavi. 2024b. Spillover effects of built-environment vulnerability on resilience of businesses in urban crises. Preprint https://doi.org/10.21203/rs.3.rs-4488088/v1.

Huang, X., Y. Jiang, and A. Mostafavi. 2023. Emergence of urban heat traps from the intersection of human mobility and heat hazard exposure in cities. *arXiv* Preprint https://doi.org/10.48550/arXiv.2301.05641.

IPCC (Intergovernmental Panel on Climate Change). 2014. *Climate Change 2014: Synthesis Report.* Contribution of Working Groups I, II, and III to the Fifth Assessment Report of the Intergovernmental Panel on Climate Change. Geneva, Switzerland: IPCC.

IPCC. 2023a. Africa. In Climate *Change 2022—Impacts, Adaptation and Vulnerability.* Working Group II Contribution to the Sixth Assessment Report of the Intergovernmental Panel on Climate Change, 1st ed. Cambridge, UK: Cambridge University Press.

IPCC. 2023b. Weather and climate extreme events in a changing climate. Pp. 1513–1766 in *Climate Change 2021—The Physical Science Basis.* Working Group I Contribution to the Sixth Assessment Report of the Intergovernmental Panel on Climate Change. Cambridge, UK: Cambridge University Press.

Jerolleman, E. M., N. Jessee, L. Koslov, C. Comardelle, M. Villarreal, D. de Vries, and S. Manda. 2024. *People or Property: Legal Contradictions, Climate Resettlement, and the View from Shifting Ground.* Springer.

Khosravi, S. 2024. Doing Migration Studies with an Accent. *Journal of Ethnic and Migration Studies* 50(9):2346–2358. https://doi.org/10.1080/1369183X.2024.2307787.

Knowledge Is Power Exhibit. n.d. My Site. https://www.preservingourplace.com/knowledge-is-power-exhibit (accessed May 28, 2024).

Lam, N. S.-N., Y. Qiang, K. Li, H. Cai, L. Zou, and V. Mihunov. 2018. Extending resilience assessment to dynamic system modeling: Perspectives on human dynamics and climate change research. *Journal of Coastal Research* 85(May):1401–1405. https://doi.org/10.2112/SI85-281.1.

Lee, C.-C., C. Chou, and A. Mostafavi. 2022. Specifying evacuation return and home-switch stability during short-term disaster recovery using location-based data. *Scientific Reports* 12(1):15987. https://doi.org/10.1038/s41598-022-20384-4.

Li, B., and A. Mostafavi. 2022. Location intelligence reveals the extent, timing, and spatial variation of hurricane preparedness. *Scientific Reports* 12(1):16121. https://doi.org/10.1038/s41598-022-20571-3.

Liu, Z., and A. Mostafavi. 2023. Collision of environmental injustice and sea level rise: assessment of risk inequality in flood-induced pollutant dispersion from toxic sites in Texas. *arXiv* Preprint https://doi.org/10.48550/arXiv.2301.00312.

Mankin, J. S., and C. W. Callahan. 2022. the scientific case for climate liability and loss and damage claims. *Lawfare.* November 14. https://www.lawfaremedia.org/article/scientific-case-climate-liability-and-loss-and-damage-claims (accessed June 24, 2024).

Mård, J., G. Di Baldassarre, and M. Mazzoleni. 2018. Nighttime light data reveal how flood protection shapes human proximity to rivers. *Science Advances* 4(8):eaar5779. https://doi.org/10.1126/sciadv.aar5779.

Masih, I., S. Maskey, F. E. F. Mussá, and P. Trambauer. 2014. A review of droughts on the African continent: A geospatial and long-term perspective. *Hydrology and Earth System Sciences* 18(9):3635–3649. https://doi.org/10.5194/hess-18-3635-2014.

Missirian, A., and W. Schlenker. 2017. Asylum applications and migration flows. *American Economic Review* 107(5):436–440. https://doi.org/10.1257/aer.p20171051.

NASEM (National Academies of Sciences, Engineering, and Medicine). 2022. *Next Generation Earth Systems Science at the National Science Foundation.* Washington, DC: The National Academies Press. https://doi.org/10.17226/26042.

NASEM. 2024a. *Climate Intervention in an Earth Systems Science Framework: Proceedings of a Workshop—in Brief.* Washington, DC: The National Academies Press. https://doi.org/10.17226/27476.

NASEM. 2024b. *Tipping Points, Cascading Impacts, and Interacting Risks in the Earth System: Proceedings of a Workshop.* Washington, DC: The National Academies Press. https://doi.org/10.17226/26925.

Niva, V., A. Horton, V. Virkki, M. Heino, M. Kosonen, M. Kallio, P. Kinnunen, et al. 2023. World's human migration patterns in 2000–2019 unveiled by high-resolution data. *Nature Human Behaviour* 7(11):2023–2037. https://doi.org/10.1038/s41562-023-01689-4.

Piemontese, L., S. Terzi, G. Di Baldassarre, D. A. Menestrey Schwieger, G. Castelli, and E. Bresci. 2024. Over-reliance on water infrastructure can hinder climate resilience in pastoral drylands. *Nature Climate Change* 14(3):267–274. https://doi.org/10.1038/s41558-024-01929-z.

Rajput, A. A., C. Liu, Z. Liu, and A. Mostafavi. 2024. Human-centric characterization of life activity flood exposure shifts focus from places to people. *Nature Cities* 1(4):264–274. https://doi.org/10.1038/s44284-024-00043-7.

Ribot, J., P. Faye, and M. D. Turner. 2020. Climate of anxiety in the Sahel: Emigration in xenophobic times. *Public Culture* 32(1):45–75. https://doi.org/10.1215/08992363-7816293.

Rigaud, K. K., A. de Sherbinin, B. Jones, J. Bergmann, V. Clement, K. Ober, J. Schewe, S. Adamo, B. McCusker, S. Heuser, and A. Midgley. 2018. *Groundswell: Preparing for Internal Climate Migration*. Washington, DC: World Bank. http://hdl.handle.net/10986/29461.

Samson, J., D. Berteaux, B. J. McGill, and M. M. Humphries. 2011. Geographic disparities and moral hazards in the predicted impacts of climate change on human populations. *Global Ecology and Biogeography* 20(4):532–544. https://doi.org/10.1111/j.1466-8238.2010.00632.x.

Schutte, S., J. Vestby, J. Carling, and H. Buhaug. 2021. Climatic conditions are weak predictors of asylum migration. *Nature Communications* 12(1):2067. https://doi.org/10.1038/s41467-021-22255-4.

Sherwood, S. C., and M. Huber. 2010. An adaptability limit to climate change due to heat stress. *Proceedings of the National Academy of Sciences* 107(21):9552–9555. https://www.pnas.org/doi/10.1073/pnas.0913352107.

Simini, F., M. C. González, A. Maritan, and A.-L. Barabási. 2012. A universal model for mobility and migration patterns. *Nature* 484(7392):96–100. https://doi.org/10.1038/nature10856.

Tschakert, P., and A. Neef. 2022. Tracking local and regional climate im/mobilities through a multidimensional lens. *Regional Environmental Change* 22(3):95. https://doi.org/10.1007/s10113-022-01948-6.

Vanos, J., G. Guzman-Echavarria, J. W. Baldwin, C. Bongers, K. L. Ebi, and O. Jay. 2023. A physiological approach for assessing human survivability and liveability to heat in a changing climate. *Nature Communications* 14(1):7653. https://www.nature.com/articles/s41467-023-43121-5.

Vargas Zeppetello, L. R., A. E. Raftery, and D. S. Battisti. 2022. Probabilistic projections of increased heat stress driven by climate change. *Communications Earth & Environment* 3:183. https://www.nature.com/articles/s43247-022-00524-4.

Verdin, A., C. Funk, P. Peterson, M. Landsfeld, C. Tuholske, and K. Grace. 2020. Development and validation of the CHIRTS-daily quasi-global high-resolution daily temperature data set. *Scientific Data* 7(1):303. https://doi.org/10.1038/s41597-020-00643-7.

Williams, E., C. Funk, P. Peterson, and C. Tuholske. 2024. High resolution climate change observations and projections for the evaluation of heat-related extremes. *Scientific Data* 11(1):261. https://doi.org/10.1038/s41597-024-03074-w.

WMO (World Meteorological Organization). 2021. *WMO Atlas of Mortality and Economic Losses from Weather, Climate and Water Extremes (1970–2019)*. WMO-No. 1267. Geneva. https://library.wmo.int/records/item/57564-wmo-atlas-of-mortality-and-economic-losses-from-weather-climate-and-water-extremes-1970-2019 (accessed June 24, 2024).

Wolde, S. G., P. D'Odorico, and M. C. Rulli. 2023. Environmental drivers of human migration in Sub-Saharan Africa. *Global Sustainability* 6(January):e9. https://doi.org/10.1017/sus.2023.5.

Xu, L., and J. S. Famiglietti. 2023. Global patterns of water-driven human migration. *WIREs Water* 10(4):e1647. https://doi.org/10.1002/wat2.1647.

Xu, C., T. A. Kohler, T. M. Lenton, J. C. Svenning, and M. Scheffer. 2020. Future of the human
    climate niche. *Proceedings of the National Academy of Sciences* 117(21):11350–11355.
    https://doi.org/10.1073/pnas.1910114117.
Zaitchik, B. F., and C. Tuholske. 2021. Earth observations of extreme heat events: Leveraging
    current capabilities to enhance heat research and action. *Environmental Research Letters*
    16(11):111002. https://doi.org/10.1088/1748-9326/ac30c0.

Appendix A

Statement of Task

The National Academies will convene a workshop to consider how an Earth systems science approach, as discussed in the 2022 National Academies' report *Next Generation Earth Systems Science at the National Science Foundation*, could be used to address climate change impacts (e.g., sustained droughts, repeated and severe flooding, increased frequency and intensity of hurricanes and cyclones, saltwater intrusion into coastal aquifers, increased risks of wildfire) and their influence on human migration. The workshop will include discussions on:

- Impacts of catastrophic events on people and infrastructure, versus chronic, sustained, and/or repeated events
- How multidisciplinary and transdisciplinary Earth systems science approaches can help advance understanding of how climate change may affect migration, improve the ability to anticipate how and where migrations might take place, and can inform strategies to address the impact of migration, especially on vulnerable populations
- Critical gaps in research and data

Appendix B

Workshop Agenda

MONDAY, MARCH 18, 2024
ALL TIMES LISTED IN U.S. EASTERN TIME

Purpose The National Academies will convene a workshop to consider how an Earth systems science approach, as discussed in the 2022 National Academies' report *Next Generation Earth Systems Science at the National Science Foundation*, could be used to address climate change impacts (e.g., sustained droughts, repeated and severe flooding, increased frequency and intensity of hurricanes and cyclones, saltwater intrusion into coastal aquifers, increased risks of wildfire) and their influence on human migration. The workshop will include discussions on:

- Impacts of catastrophic events on people and infrastructure, versus chronic, sustained, and/or repeated events
- How multidisciplinary and transdisciplinary Earth systems science approaches can help advance understanding of how climate change may affect migration, improve the ability to anticipate how and where migrations might take place, and can inform strategies to

address the impact of migration, especially on vulnerable populations
- Critical gaps in research and data

9:30	Registration

10:00–10:15   **Welcome and Opening Remarks**
Amir AghaKouchak,* University of California, Irvine, Organizing Committee Chair
Alexandra Isern, National Science Foundation

10:15–11:15   **Keynote: Broad Overview on Climate Change and Human Migration: Current Status and Research Gaps**
Justin Mankin, Dartmouth College
Lori Hunter,* University of Colorado Boulder
Jackie Qataliña Schaeffer, Alaska Native Tribal Health Consortium
Moderator: Kilaparti Ramakrishna,* Woods Hole Oceanographic Institution

11:15–12:30   **Session 1: Mechanisms and Pathways for Modeling the Impacts of Catastrophic Events on Human Migration**
Giuliano Di Baldassarre, Uppsala University
Nina Lam, Louisiana State University
Serena Ceola, University of Bologna
Elizabeth Marino, Oregon State University Cascades
Moderator: Elizabeth Fussell,* Brown University

12:30–1:15   Break

1:15–2:30   **Session 2 Mechanisms and Pathways for Modeling the Impacts of Chronic and Slow-Onset Events on Human Migration**
Kevin Anchukaitis, University of Arizona
Cristina Cattaneo, CMCC Foundation—Euro-Mediterranean Center on Climate Change, Italy and RFF-CMCC European Institute on Economics and the Environment, Italy
Cascade Tuholske, Montana State University
Moderator: Shanna N. McClain,* National Aeronautics and Space Administration

2:30–2:40   Break

2:40–3:55   Session 3: Transdisciplinary Research and Collaborations
                 Pierre Gentine, Columbia University
                 Katharine Donato, Georgetown University
                 Diana Liverman, University of Arizona
                 Moderator: Diego Pons, University of Denver

4:00–4:15   Day 1 Recap
                 Amir AghaKouchak, University of California, Irvine

                 END OF DAY 1

                 TUESDAY, MARCH 19, 2024

10:00–10:10   Welcome
                 Amir AghaKouchak,* University of California, Irvine,
                     Organizing Committee Chair

10:10–11:25   Session 4: Regional Versus Global Perspectives for
                 Modeling Climate Change–Migration Impacts
                 Paolo D'Odorico, University of California, Berkeley
                 Alexander de Sherbinin, Columbia University
                 Meda DeWitt, The Wilderness Society
                 Moderator: Jackie Qataliña Schaeffer,* Alaska Native
                     Tribal Health Consortium

11:30–12:45   Session 5: Harmonization of Spatial and Temporal
                 Data and Models to Advance an Earth Systems Science
                 Approach to Modeling Climate Migration
                 Chris Funk, University of California, Santa Barbara
                 Ali Mostafavi, Texas A&M University
                 Deborah Balk, City University of New York
                 Moderator: Somayeh Dodge,* University of California,
                     Santa Barbara

1:15–2:00   Closing Remarks
                 Danielle N. Poole,* Yale School of Public Health
                 Shanna N. McClain,* National Aeronautics and Space
                     Administration
                 Elizabeth Fussell,* Brown University
                 Laura Lautz, National Science Foundation

2:00   MEETING ADJOURNS

Appendix C

Workshop Planning Committee Biographies

Amir AghaKouchak is a professor of civil and environmental engineering and Earth systems science at the University of California, Irvine. His research focuses on natural hazards and climate extremes and crosses the boundaries between hydrology, climatology, and remote sensing. One of his main research areas is studying and understanding the interactions between different types of climatic and nonclimatic hazards, including compound and cascading events. He has received several honors and awards, including the American Geophysical Union's James B. Macelwane Medal and the American Society of Civil Engineers Huber Research Prize. AghaKouchak is currently the editor in chief of *Earth's Future*, a transdisciplinary scientific journal examining the state of the planet and the science of the Anthropocene. He received a B.S. and an M.S. in civil engineering from the K.N. Toonsi University of Technology in Tehran, Iran, and a Ph.D. in civil and environmental engineering from the University of Stuttgart, Germany. AghaKouchak recently served on the National Academies' Workshop Planning Committee on Tipping Points, Cascading Impacts, and Interacting Risks in the Earth System.

Somayeh Dodge is an associate professor of spatial data science in the Department of Geography at the University of California, Santa Barbara (UCSB). Before joining UCSB in 2019, she served as an assistant professor at the University of Minnesota (2016–2019) and the University of Colorado, Colorado Springs (2013–2016), and was a postdoctoral fellow at The Ohio State University (2012–2013). Dodge's research focuses on developing data analytics, knowledge discovery, modeling, and visualization techniques to study movement and behavioral responses to environmental disruptions across multiple scales in human and ecological systems. She is a recipient of the 2021 National Science Foundation CAREER award and the 2022 Emerging Scholar Award from the Spatial Analysis and Modeling Specialty Group of the American Association of Geographers. She currently serves as a director on the Board of Directors of the University Consortium for Geographic Information Science, the co-editor in chief of the *Journal of Spatial Information Science*, and a member of the editorial boards of multiple journals, including *Geographical Analysis, Cartography and Geographic Information Science*, and the *Journal of Geographical Systems*. Dodge received a Ph.D. in geography with a specialization in geographic information science from the University of Zurich, Switzerland.

Elizabeth (Beth) Fussell is a professor of population studies and environment and society at Brown University. She has investigated the long-term effects of Hurricane Katrina on the residential mobility, health, and well-being of the residents of New Orleans using innovative methods and data sets. Fussell has extended this research to study the effects of hurricanes and other exogenous shocks on individual and household migration and internal migration systems in the United States, with a new focus on Puerto Rico. She gave the keynote to the joint meeting of the National Academies' Mapping Science Committee and Geographical Sciences Committee, Humanitarian Responses to Forced Migration and Displacement: New Insights from Quantitative and Qualitative Geographic Data. Fussell has presented to the National Academies' Gulf Research Program and the National Academies' Committee on Population and participated in several National Academies' meetings, all of which focus on leveraging data for and measuring key concepts in disaster research. She received her Ph.D. and M.A. from the Department of Sociology at the University of Wisconsin–Madison, concentrating in demography, and was a National Institutes of Health Postdoctoral Fellow at the University of Pennsylvania Population Studies Center.

Lori Hunter is the director of the Institute of Behavioral Science at the University of Colorado Boulder, where she is also a professor of sociology. Hunter's research and teaching focus on links between environmental context and human population dynamics. Specific settings include rural South Africa and Mexico, where her scholarship connects rural livelihood strategies, including migration, to local shifts in rainfall, temperature, and natural resource availability. She has been an invited speaker on the topic of migration and climate change at a variety of settings, including the United Nations, the National Academies, the Rio+20 Earth Summit, Future Earth, and the French Institute for Demographic Studies. Hunter received her

Ph.D. from Brown University. She is a member of the National Academies' Board on Environmental Change and Society and the Board's liaison to the National Academies' Committee on Managed Retreat in the U.S. Gulf Coast Region. Hunter is a member of the National Academies' Roundtable on Macroeconomics and Climate-Related Risks and Opportunities.

Shanna N. McClain is the disasters program manager for the National Aeronautics and Space Administration's (NASA's) Earth Science Applied Sciences Program. She and her team promote the use of Earth observations to support decisions made across the disaster cycle. McClain endeavors to define new and innovative opportunities for applying Earth science information through the development of partnerships and projects in fragile and crisis-affected communities to build a more risk-informed global society. She joined NASA after working with the Environmental Law Institute to co-lead policy development and programming on environmental and climate-related migration and displacement and programming on environmental conflict and peace. McClain previously held consultancies with the United Nations Environment Programme/Office for the Coordination of Humanitarian Affairs Joint Environment Unit focused on the integration of environmental considerations in sudden-onset and protracted humanitarian crises, including developing guidelines for how to prepare for and respond to technological, industrial, and nuclear disasters. McClain was awarded a National Science Foundation Integrative Graduate Education and Research Traineeship fellowship, which provided her the opportunity to work with the International Commission for the Protection of the Danube River in Vienna, Austria. She was also awarded the American Association for the Advancement of Science's Science and Technology Policy Fellowship, which offered the opportunity to contribute evidence-based scientific knowledge and skills to the development of federal government policies and actions on disaster resilience. She earned a Ph.D. in environmental resources and policy from Southern Illinois University.

Diego Pons is a research associate scientist at the University of Denver. He is an applied climatologist with a background in biology and paleoclimatology. Pons taught physical geography at Colorado State University after spending the last 2 years as a postdoctoral research scientist fellow at Columbia University. At Columbia University, he worked at the intersection between climate and society co-developing seasonal and subseasonal forecast systems to support decision-making processes at the farm level in rain-fed agricultural landscapes in Latin America. Pons's research in mountainous regions of Mesoamerica has been funded by the National Science Foundation, including climate reconstructions in tropical settings using tree rings and assessing the relationship between climate and migration. He uses the climate information at different timescales (from months to millennia) to investigate historical and future human–environment interactions and the implications for policy making and development. Pons's research interests are dendrochronology, climate variability and change, seasonal to subseasonal climate forecasts, climate impacts on agriculture, climate–health–migration interactions, and satellite-derived vegetation monitoring. Pons graduated from the University of Denver as a geographer funded by a Fulbright scholarship.

Danielle N. Poole is an associate research scientist at the Yale School of Public Health. She is a population health scientist notable for her contributions to the evidence base for decision making in conflict- and disaster-affected settings. Within the broader field of humanitarian health research, her work is centered around measuring and responding to the health needs of populations in transit and adapting and developing novel research methods for complex settings. Poole's transdisciplinary approach provides evidence of the dynamic interaction of individual, social, structural, and "place" determinants of health. Her findings have been featured in numerous academic meetings, including at the National Academy of Sciences, and journals, including *The Lancet* and *Nature*. Poole is a member of the International Organization for Migration's Migration Health and Development Research Initiative, has served as a fellow with the United Nations' Office for the Coordination of Humanitarian Affairs, and has led independent research with numerous intergovernmental agencies, including the World Health Organization and the World Food Programme, as well as nongovernmental organizations. She holds a Ph.D. from Harvard University, an M.S.P.H. from Brown University, a B.S. from Seattle University, and completed postdoctoral training at the Neukom Institute for Computational Science at Dartmouth College.

Jackie Qataliña Schaeffer, an Iñupiaq from Kotzebue, Alaska, is the director of the Climate Initiatives program at the Alaska Native Tribal Health Consortium under the division of Community Environment and Health. For decades she has worked across Alaska holistically infusing Indigenous knowledge into a variety of sectors she has experience in, including comprehensive planning, energy, housing, water security, sanitation, and climate change adaptation for rural communities. Her passion is to serve the Indigenous people of Alaska and provide an Indigenous perspective to all her work, including the importance and recognition of traditional philosophies, knowledge, and worldviews. Her current work includes climate change assessments, community engagement, community relocation oversight, and overseeing the Center for Climate and Health and the Center for Environmentally Threatened Communities within the Climate Initiatives program. Qataliña has co-authored six regional Energy Plans for the State of Alaska, the Oscarville Tribal Adaptation Plan, 2019, and currently works with Newtok and Kivalina on community relocation due to climate change. She is a co-principal investigator on the Human Well Being team for the Study of Environmental Arctic Change, a National Science Foundation–funded project through the University of Alaska Fairbanks.

Kilaparti (Rama) Ramakrishna is the director of the Marine Policy Center and a senior advisor to the president on ocean and climate policy at the Woods Hole Oceanographic Institution. Prior to this, he worked extensively with the United Nations (UN) as head of strategic planning at the Green Climate Fund; head of the Economic and Social Commission for Asia and the Pacific: Subregional Office for East and North-East Asia (ESCAP-ENEA) Office, covering six member states of ESCAP (China, Democratic People's Republic of Korea, Japan, Mongolia, Republic of Korea, and Russian Federation) and two associate members (Hong Kong and Macao); as chief of cross-sectoral environmental issues and principal policy advisor to the executive director of the UN Environment Programme. Ramakrishna also provided secretariat services to the North-East Asian Subregional Programme for Environmental Cooperation and was a lead author of the Fifth Assessment (and many before it) by the Intergovernmental Panel on Climate Change; coordinating lead author of the Millennium Ecosystem Assessment; and lead author of the Interlinkages Assessment. Before joining the United Nations, Ramakrishna worked for many years as director of science in public affairs and the vice president at the Woods Hole Research Center in Massachusetts. During this time, he taught at several law schools, including at the Fletcher School of Law and Diplomacy, Harvard Law School, Boston University and Boston College law schools, and Brandeis and Yale Universities. Ramakrishna is an elected life member of the U.S. Council on Foreign Relations. He is also the chair of the Strategic Advisory Group of the Nippon Foundation-GEBCO Seabed 2030 Project; a member of the Advisory Board of Back to Blue, a global initiative of Economist Impact; and a member of the Board of Directors of the Woodwell Climate Research Center and the Consensus Building Institute. Ramakrishna holds a B.Sc. and a B.L. in sciences and law and an M.S. and a Ph.D. in international law.